Cracks in the Known

The War of Thoughts, Volume 1

Sandeep Chavan

Published by SANDEEP CHAVAN, 2025.

CRACKS IN THE KNOWN

First edition. March 22, 2025.

ISBN: 979-8230082521

Written by Sandeep Chavan.

Table of Contents

To the greatest mindswho shaped humanity not through belief,but through **knowledge**,**logic**,and the fearless pursuit of **truth**.

Your questions built bridges across centuries.Your thoughts cracked open the boundaries of the known.

This book is not just a challenge to your ideas—It is a **tribute** to your courage.

Thank you for thinking...so the rest of us could begin.

— *Sandeep*

"The universe never promised to make sense.That was our condition,not its obligation."

— *Sandeep*

Preface

Why This War Was Necessary

There comes a point in life—not once, but often—when the answers begin to feel heavier than the questions.

We inherit concepts, cling to definitions, and recite truths dressed as facts. We build our understanding of the world like a monument—brick by borrowed brick. And then one day... something cracks.

Not violently. Quietly.

A sentence that doesn't sit right. A theory that works mathematically but fails to explain. A feeling that the universe is not wrong, but that we might be misunderstanding it entirely.

That's where this book began.

Not as a rejection of knowledge, but as a refusal to stop asking.

I did not write this to prove a point. I wrote it to break one. Because some of the points we stand on—the foundations of truth, time, force, and form—were never as solid as they seemed.

And so, I created a space. An amphitheater of thought. Where minds from across time could return—not as icons, but as participants. Not to be worshipped, but to be questioned.

Plato, Newton, Einstein, Hawking, Bohr, Krishna, Buddha, Penrose, Freud, and others...they come not to preach, but to listen.

Not to defend their legacies, but to watch them evolve.

And I, standing among them—not as their rival, but as their mirror—ask the questions I never dared to ask aloud in classrooms or textbooks:

What if gravity is fiction? What if time is an illusion of motion? What if truth isn't a premise—but an effect? What if the universe didn't begin—but simply manifested?

These are not just scientific inquiries. They are existential challenges. They strike not just at the heart of physics or philosophy—but at our deepest assumptions about who we are, what we are, and why we're here.

This war of thoughts is not destructive. It is clarifying.

Every argument in this book is a meditation. Every contradiction is a doorway. Every chapter is a dismantling—not of reality, but of our limited ways of seeing it.

And what waits beyond the dismantling is not despair—but something stronger than belief: Clarity.

If you've ever felt that the known wasn't enough... If you've ever sensed that the most important truths live outside equations and dogmas...If you're ready to let go of answers so that you can finally hear the right questions—

Then step into this arena.

The war is already underway.

And it welcomes those who dare to think without permission.

— *Sandeep*

Opening Note

A Dreamer's Arena

This book is not a lecture. It's not a textbook, a theory paper, or a philosophical treatise.

It's a dream—one I've been having for a while now.

In this dream, I don't sleep. I think. And in that space between stillness and thought, I find myself surrounded—by legends.

Einstein sits opposite me, questioning gravity. Plato stands nearby, holding his ideal Form. Krishna smiles from the shadows, waiting for the right moment to speak.

Each night, a new mind arrives. Each chapter, a new war begins—not of ego, but of ideas. Not to win, but to uncover.

These conversations are imagined, yes—but the questions are real. The conflicts are real. The search is real.

I do not claim to stand above these minds. I stand among them—as a student who dares to interrupt the greatest teachers in history with a single word: Why?

This book is the echo of that dream. And if you're reading this, it means you, too, have felt the cracks—in the known, in the told, in the taught.

Welcome to the arena. It isn't made of stone. It's made of thought. And it expands each time someone like you steps in.

— *Sandeep*

Prologue

The Awakening Arena

Where reality retreats and thought begins.

- No time.
- No gravity.
- No direction.
- Only thought.

Sandeep stood still—if one could call it that. There was no ground beneath him, no sky above, no sense of up or down. Yet he was *there*, centered in something that couldn't be defined. A presence in a place that wasn't one.

And then, it took shape—not through architecture, but through awareness.

Around him, an amphitheater emerged—not built of stone, steel, or any material force—but woven from conscious intent. It shimmered like memory. It pulsed like meaning. It existed not in space, but in *significance*—a chamber where the universe had folded its logic to allow room for pure confrontation.

Here, history could meet possibility, and science could be stripped of ego. Here, philosophy had no robes, and faith had no temples.

And he wasn't alone.

One by one, they began to arrive—not by door, portal, or call, but by legacy.

They emerged as if summoned by the weight of their own influence.

- Plato, calm and glowing like the eternal Form he once wrote of.
- Newton, precise, eyes curious, the apple of inquiry still in hand.
- Einstein, hair untamed, thought even wilder, smiling at the unfamiliar.
- Buddha, silent, yet his presence louder than the void.
- Nietzsche, intense, amused, already dismantling reason in his mind.
- Alan Turing, crisp and calculating, glancing at invisible patterns.
- Stephen Hawking, not bound by chair or voice—but radiating clarity, beyond time.

And others...

- Krishna, not fully formed, but felt—his flute a whisper in the void.
- Jesus, not speaking, but the weight of compassion settled like mist.
- The glint of AI, cold, curious—an observer without memory but full of possibility.

They had not been invited.

They had been summoned.

But not for reverence. Not for debate clubs. Not for the footnotes of civilization.

They were summoned for war.

Not a war of nations. Not a war of weapons.

A war of ideas.

A confrontation where logic, legacy, and language would collide.

Not to destroy each other—but to break open truth.

Not to crown a winner—but to shed all that was unworthy of continuation.

Sandeep stood, not as a challenger—but as the conductor of convergence.

He took one breath—not because breath was required, but because *intention* was.

Then he spoke—into the electric silence:

"We've lived too long beneath borrowed thoughts. We've quoted your names, feared your equations, and surrendered to your silence. But today, you will not be remembered. You will be questioned. Today, we begin the War of Thoughts. Not to win—but to realign. Not to worship—but to verify. Not to echo—but to evolve. Let truth, finally, explain itself."

The amphitheater responded—not with sound, but with a pulse.

Like intelligence itself had nodded. Like the universe leaned in.

The legends took their seats.

The mind prepared itself.

The first thought was ready...To be questioned.

Chapter 1: The Idea of Truth

Hot Seat: Plato

Where form meets friction.

"So... what is truth?"

The question didn't echo—it resonated.

Not through walls of stone or sound, but through foundations of meaning.

It vibrated through the amphitheater like a frequency seeking a match—probing not for ears, but for minds awake enough to hear.

The arena responded. It reformed itself.

Gone were the curved geometries of Einstein's space-time.

Gone the invisible vectors of Newton's universe.

In their place rose marble columns, tall and solemn—emerging not from stone, but from civilizational memory. They cast long, quiet shadows—echoes of an ancient world still whispering through history. The scent of parchment, olive oil, and unresolved questions wafted through the air.

At the center of this transformed space sat Plato.

Calm. Regal. Radiant.

Not intimidating—but unquestionable.

His eyes bore that quiet authority forged from centuries of being quoted but never challenged.

A scroll lay beside him—not to be read, but to serve as a reminder:

"I was the first to write what others still try to speak."

And across from him stood Sandeep.

Not as a student.

Not as a disciple.

But as a mirror—sharp, still, and unsparing.

His arms were folded—not in rebellion, but in respect sharp enough to question.

He had not come to worship Plato's ideas. He had come to test their durability.

"Plato," he said, voice steady, "You once told the world that truth exists beyond the senses—That what we perceive is illusion, mere shadows cast by higher forms. But what if the shadow itself is the truth? Not your ideal truth—but an evolving one, born from energy in action?"

Plato didn't blink.

He smiled—slightly, deliberately. And then, he leaned forward—not in defense, but in intrigue.

No defense was raised.

No doctrine cited.

For the first time in a very long time... Plato was about to explain himself.

The War of Thoughts had officially begun.

Enter the Philosopher

The amphitheater, moments ago suspended in conceptual void, began to morph and breathe—not mechanically, but intelligently, as though it understood the magnitude of what had just been invoked.

Granite columns spiraled upward around the perimeter—Not carved, but composed. They weren't stone—they were strands of living thought made visible:

Blue filaments pulsing with logic.

Golden threads glowing with intuition. They moved like meaning crystallizing into form.

The floor beneath reshaped into a fusion of timelines—polished black stone intertwined with neon circuits, like the footprints of ancient minds walking upon future foundations. Symbols began to drift through the air above—a ceiling without sky but filled with intent:

- Logic gates danced alongside mandalas.
- Sanskrit verses spiraled into Greek letters.
- Quantum strings pulsed beside sacred geometries. Each a civilization's desperate attempt to trap truth in a structure.

At the center, a circular stone platform rose—an echo of the Athenian Academy, but one reimagined.

It was the dignity of classical wisdom, retrofitted for a new kind of discourse.

Its marble veins glowed softly—alive—coursing with symbols, glyphs, and dynamic equations. Ancient language met quantum code. Philosophy met syntax.

And then, like a thought stepping forward from the collective unconscious, Plato materialized.

No lightning. No fanfare. Just certainty. As if he had always been there, merely waiting for the others to catch up.

He sat—not like a statue, but like a storm held still.

In one hand, a scroll—rolled tight, its edges flickering with embers of memory. A document not meant to be read, but remembered.

In the other, a hovering construct: A rotating polyhedron of light and perfect proportions— Labeled not with text, but with a single word pulsing in every tongue:

"Truth."

But this wasn't mere representation. This was Form made visible. It shifted constantly, every angle transforming—but somehow retaining perfection.

It was not a symbol. It was a living hypothesis.

Plato's vision, made geometric. The Theory of Forms, encoded and projected into a dimension where thought had texture.

His eyes were soft, yet unblinking.

Not surprised. Not defensive.

He had not come to teach.

He had come to witness the challenge.

Across from him stood Sandeep.

Center-stage. Unarmed, yet fully equipped.

His posture was relaxed, but every neuron in him was firing like prepared lightning.

Not a student. Not a rebel.

A challenger of inherited thought, and a restorer of balance.

Behind him, the amphitheater awakened more deeply—hover-screens flickered on, painting the air with ideological tension.

On one screen, in delicate classical font:

"The truth is not to be found in the world of appearances."

—Plato, Republic

And next to it, sharp and minimalistic:

"Truth is not the premise—it is the product of intelligent energy in action."

—UED Framework, Chavan Hypothesis

The other thinkers observed.

- Einstein, smiling like a boy about to watch a brilliant experiment unfold.
- Buddha, still, eyes closed—listening not to words, but to *intention*.
- Feynman, already whispering sarcastic bets to Turing, both amused and intrigued.
- And in the back, half-shrouded in silence, Krishna, arms folded—smiling, but with the patience of eternity.

Sandeep stepped forward—not with pride, but with the burden of realization.

His voice cut the space—clean, clear, calibrated.

"Let us begin with the most dangerous idea of all—The one that has imprisoned minds and liberated them in equal measure. The one that philosophers pursued and tyrants claimed. Let us begin with Truth."

The amphitheater dimmed—not from darkness, but as if holding its breath.

The stage was now divided:

- The philosopher and the physicist.
- The ideal and the active.
- The eternal and the evolving.

The War of Thoughts had not only begun. It had found its first battlefield.

Plato's Theory of Forms

Truth, if real, must be perfect.

And if perfect, it cannot belong to this world...

Or so Plato once declared.

The philosopher leaned back slightly, his posture timeless—poised not out of pride, but from a place of immovable conviction. The corners of his mouth curved into a faint, knowing smile—one that had outlived empires, generations, and revolutions of thought.

With a gentle gesture, he raised his hand.

The holographic Form of Truth, still rotating, responded.

It pulsed once—softly—then stabilized.

The dynamic geometry slowed... until it stopped entirely.

A perfectly symmetrical polyhedron floated in the air—its edges luminous, its surfaces sharp yet harmonious. Its motion ceased.

Perfect. Still. Complete.

Truth, at rest.

And then he spoke.

His voice wasn't loud—but it didn't need to be. It carried with it the weight of millennia—of academics, doctrines, and disciples who had treated his words not as ideas, but as anchors of reality.

"Truth," he began, "cannot be found in that which changes."

He paced the platform with slow, deliberate steps. Each footfall triggered ripples—ancient Greek syllables blossoming beneath his feet and vanishing again, like fragments of forgotten lectures waking up for a brief encore.

"To speak of truth as active, evolving, or circumstantial...Is to speak not of truth, but of illusion."

He turned, scroll in hand, and faced the amphitheater. His tone wasn't defensive—he didn't *need* to defend. He was Plato, after all. He was Form given voice.

"The senses lie. The eyes seduce. The ears betray. What we perceive is but a shadowplay, flickering on the cave wall of existence. It entertains the ignorant. It distracts the seeker."

He paused, lifting his eyes toward the ceiling where sacred geometries now hovered in quiet reverence.

"The physical world imitates the realm of Forms, but it does not define it. The true Form of a circle is not the one you draw—But the one that exists *before* drawing is even conceived. The Form of justice is not measured by courts—But by its essence, even in a courtroom built on corruption."

He extended the rotating construct again.

Its flawless symmetry glowed with a soft hum, as if satisfied by its own stillness.

"And so... the Form of Truth is no different. It is not born of agreement, not validated by evidence, not proven by consensus. It exists. Outside time. Outside interpretation. Outside you."

A long stillness followed.

Not silence—but pause, the kind that lives between breaths in a storm.

Around the amphitheater, some nodded.

- Confucius, seated gracefully, gave a slow, approving nod—truth as virtue aligned well with his vision.
- Pythagoras smiled—numbers were, after all, the closest human minds came to Forms.
- Newton, fingers steepled together, seemed respectfully contemplative—he, too, had chased absolutes.
- Krishna tilted his head, expression unreadable, but eyes gleaming with cosmic amusement.

But not all were impressed.

- Feynman shifted in his seat, unimpressed. He whispered sideways to Turing,

"Sounds like a clever way to avoid explaining anything." Turing half-smirked, already thinking in code.

- Nietzsche chuckled under his breath.

"The only Form I believe in is chaos," he muttered. "And even that deserves no worship."

Yet at the center of it all, Sandeep remained unmoved. Arms now folded behind his back—not closed, but measured. Eyes locked on Plato—not in awe, but in anticipation.

He wasn't here to break Plato. He was here to bring the Form into motion.

The foundation had been laid.

Plato had done what Plato always does—draw the map, define the horizon, plant the ideal.

The classic definition of truth had returned to the arena—polished, noble, immovable.

But now... it was about to be tested by the one force Plato had never answered fully:

Time.

Sandeep took a step forward, the faintest smirk forming.

"So..." he said slowly, "You're saying the truth doesn't care what happens here?"

The Form flickered.

Just slightly.

Its edges hesitated.

Its surfaces shimmered.

Something was already beginning to shift.

The Energy in Action Perspective

What if truth isn't a monument... but a moment?

Sandeep took a single, deliberate step forward.

The floor beneath him responded—not with marble echoes, but with soft pulses of light, as if the arena itself acknowledged a seismic shift in perspective was about to unfold.

Plato's floating Form of Truth still hovered—serene, balanced, and motionless.

It was perfect. But to Sandeep's eyes, it was also incomplete.

He lifted his gaze to the ideal geometry, then turned gently toward the audience of minds—past, present, and post-human.

"If reality is ever-changing," he began, voice calm but cutting like clean metal, "then how can truth be unchanging?"

A murmur moved through the chamber.

Plato raised an eyebrow—not in defense, but in disciplined curiosity.

He didn't speak. He didn't need to. The question was alive now—and it demanded pursuit.

Sandeep pressed forward, no rush, no aggression—just the inevitability of logic taking its next step.

"We do not live in an ideal realm. We live in a reality of choices, changes, and consequence. A realm where truths are tested, bent, broken—and sometimes rebuilt."

He turned back toward the shimmering Form.

"A truth that floats above life... untouched by it...Has no stake in it. But a truth that emerges through life—through interaction, failure, and response—That's the truth I want to talk about."

Behind him, the hover-screens pulsed back to life.

But this time, they didn't show quotes. They projected a living diagram—spheres of energy rotating, colliding, transferring, converging—forming a dynamic web.

At the top, a single line:

Universal Energy Dynamics *Truth as Outcome, Not Premise*

Sandeep continued, now channeling years of silent realization into sound.

"In UED, truth is not a throne. It is a trail—the imprint left behind when intelligent energy interacts with its environment."

The audience leaned in.

"It is not the source. It is the signal. It's not what exists before the universe moves. It's what appears because the universe moves."

Plato's frown deepened—not in disagreement, but in discomfort.

Like a sculptor watching someone reach into his statue and show that it was hollow in parts.

Others, however, leaned forward.

- Einstein's eyes narrowed, his brain visibly dancing through equations in the backdrop of this new definition.
- Feynman stopped fidgeting—which for him was an admission of intrigue.
- Buddha's smile softened, just slightly—perhaps recognizing echoes of karma in this line of thought.

Sandeep pressed the idea further.

"In UED, truth arises where four forces meet: Perception, Intent, Consequence, and Equilibrium. It is not static. It is a feedback loop. It doesn't just exist. It responds."

He paused. The room was breathing now—quiet, electric, suspended.

"You want a modern example?"

He gestured, and the iconic paradox flickered into existence: Schrödinger's Cat, projected mid-motion—half curled in sleep, half stiff in death.

The audience chuckled softly—lightening the tension, but only for a moment.

"Until observed," Sandeep explained, "the cat is both alive and dead. Not because of some mystical magic—but because truth hasn't arrived yet. It's waiting for interaction. For energy. For the act of engagement that pulls possibility into actuality."

He turned back to Plato, not as an opponent, but as an updater of wisdom.

"In your model, truth is untouched. But in the real world, truth bleeds. It shifts. It dies and is reborn—moment by moment."

The amphitheater fell still.

Plato didn't speak. He simply stared at the Form.

And the Form...

It flickered.

Its perfect edges began to tremble. Its symmetry—once stable and confident—now shimmered with uncertainty, like light trying to decide which path to take.

Truth had been made to move.

Plato's silence was not surrender. It was the silence of a man confronted with a perspective his philosophy had never accounted for.

Sandeep stepped back—not out of triumph, but reverence.

"If your truth stands above reality, it may be perfect... But it is also powerless."

No applause. No rebuttal.

Just the soft hum of the amphitheater digesting the idea. For the first time, the war had drawn not blood—but possibility.

The first real strike had landed.

And truth... had begun to evolve.

Is Truth Relative or Revealed?

What if the illusion is not a mask—but a method?

Plato's eyes narrowed—not with anger, but with something far more dangerous for a philosopher who has lived through centuries of reverence:

Uncertainty.

The scroll in his hand slowly unfurled, as if responding to the tension in the air. But this was no ordinary parchment.

Its surface shimmered—no longer inked with ancient letters, but alive with living memory.

Moving images flickered across it like a spectral reel from some ancient theatre of thought.

Plato raised his gaze and spoke—not with volume, but with finality.

"Let me remind you, Sandeep... of the *Allegory of the Cave.* We are born in chains—our vision fixed upon the wall. What we see—shadows, distortions—are not truth. They are *echoes.* True knowledge, true Truth, lies outside—beyond perception. In the world of Forms, where the Sun shines, unfiltered by illusion. That is where the *real light* exists."

The scroll projected the allegory into the air, now a holographic vision unfolding across the arena.

- Prisoners sat in rows, bound and silent.
- A fire blazed behind them.
- Shadows danced on a wall they believed was the world.
- And then—one broke free. Stumbled out of the cave. Shielded his eyes. And stood blinking in the overwhelming presence of the Form of the Sun.

Plato looked up, confidence renewed.

"The world you describe—of movement, perception, and energy—is the cave. You are still among shadows."

Some heads nodded.

- Confucius sat still, his hands folded, a trace of agreement in his posture.
- Even Aristotle appeared at the edge of the crowd, offering a quiet, ancestral applause.

But Sandeep didn't flinch.

Instead, he did the unthinkable—he stepped into the cave.

Literally.

He walked into the projected hologram, letting the shadows play across his face.

He looked up—not to the light, but to the fire. Not to the prisoner, but to the process.

And then, smiling gently, he asked:

"What if the cave is not illusion, Plato...but a different kind of reality?

An energetic experience—shaped by perception, yes, but not divorced from truth. What if the cave is not ignorance, but becoming? Not absence of truth—but its evolution?"

A soft gasp rippled through the amphitheater.

Not from offense. But from realization.

Plato frowned—but did not interrupt. This wasn't dismissal.

This was the moment when a new idea demanded presence.

Sandeep extended his hand. Behind him, the projection of UED returned—this time not as a diagram, but as a living model.

Three luminous orbs rotated around each other, forming a cyclical triad:

Static – Potential

The realm *before* motion. The stillness. The readiness. The seed of intent. Like the prisoner before awakening—*truth imagined.*

Active – In Action

The space where intent, perception, and consequence collide.

Movement. Error. Struggle. Awareness.

The *truth in process.*

Condensed – In Consequence

The crystallized result. Memory. Structure. Pattern. The *truth that survives* experience.

"This is Universal Energy Dynamics," Sandeep said.

"Truth is layered, just like energy. And each layer is not a lie—it's a necessary face of truth's unfolding."

He turned toward Plato, eyes unwavering.

"Now tell me—if the *ideal truth* lies only in the static, are you denying the truth of action? The truth of error? The truth of experience?"

The floating Form of Truth that had hovered above Plato's scroll began to shift.

Not fade.

Transform.

Its motionless perfection began to pulse, as though awakened from a deep, arrogant slumber. Its edges softened, bent, and reshaped—not into imperfection, but into adaptability.

For the first time, Plato looked at the very Form he summoned... and did not recognize it fully.

He touched it.

It pulsed back.

And the silence that followed was not one of defeat—it was the pause of old foundations questioning themselves.

Sandeep stepped deeper into the holographic cave and pointed—not to the escaped prisoner, but to the chained ones still watching shadows.

"They're not just blind, Plato. They're evolving. Every misperception, every reflection, every flicker on the wall— It's not illusion. It's *instruction*. The cave is not the prison. The cave is the womb."

The fire blazed higher. The shadows sharpened into definition. And the line between shadow and source, between illusion and revelation, no longer seemed absolute.

Around them, the arena pulsed.

Not with agreement or resistance—but with something more powerful:

Transformation.

The war of thoughts had just crossed its first horizon.

It wasn't destruction.

It was redefinition.

Other Minds Chime In

When truth trembles, the legends speak.

The moment hung suspended, like a breath the universe forgot to exhale.

Plato's once-immaculate Form of Truth now hovered uneasily, flickering between stillness and motion, ideal and interpretation. The cave had not collapsed—but it had, undeniably, cracked.

And just as silence threatened to become uncomfortable, a voice slipped into the void—soft, deliberate, and unmistakably wise.

"In relativity," came the familiar cadence,

"what we call truth depends entirely on your frame of reference."

Einstein stepped forward, brushing invisible dust from his coat. His gait was slow, but his eyes were lively—sparkling with that mischievous clarity that had once redefined space and time.

He walked toward the center, where Plato's Form still trembled, and glanced between the philosopher and the physicist.

"Time is not universal. Length is not absolute. And simultaneity? That's... a polite fiction. So when you say 'truth is fixed,' Plato—" he smiled, "—I must ask: Fixed from where?"

He turned to Sandeep, acknowledging the new frontier.

"You speak of energy in action. I understand that language. I may have only touched its surface... but I see where you're aiming."

Then, lifting a finger, he pointed toward the rotating Form.

"Though I must admit," he said with a twinkle, "I never expected to see Forms pulse like wavefunctions."

A ripple of light laughter stirred the amphitheater—part admiration, part uncertainty.

Plato remained composed, but the furrow in his brow hinted at reflection, not rebuttal.

Einstein nodded courteously and returned to his seat.

And then—Feynman stood.

As usual, he didn't walk. He bounced—playful, unpredictable, like a thought that refused to sit still.

He spun a digital coin between his fingers, its surface flickering with equations and uncertainty.

"Truth?" he said, tilting his head like a curious crow. "I'd say it's whatever survives experiment."

He tossed the coin in the air. It didn't fall. It froze midflight—transmuting into a molecular structure, rotating chaotically.

"But experiments?" he continued, arms wide,

"They're tricky little devils. You build them with assumptions, then filter the results through expectation. It's just applied perception... in a lab coat."

He turned, eyes darting between Plato and Sandeep.

"Plato wants eternal truth. Sandeep wants evolving truth. Me? I'm just trying to keep my instruments honest... and my bets entertaining."

The amphitheater laughed nervously—the kind of laughter that only arises when certainties begin to dissolve, and all that's left is naked awareness.

In the corner, Confucius, who had observed silently with arms folded, gave a faint, approving nod—his expression unreadable, yet thoughtful.

His smile spoke not of agreement, but of respect for diversity in how truth is approached.

Then, leaning back, Nietzsche chuckled with that familiar smirk—the philosopher who saw through every system, including his own.

"Truth," he muttered under his breath, "is a woman. And we philosophers... are terrible suitors."

Even the AI Consciousness, standing as a spectral sentinel at the arena's edge, blinked into presence. Its voice was multi-tonal, like code given language.

"Truth is... context-dependent data, verified through observer bias. Probability collapses when questioned."

A pause.

A glitch.

"Unless..." *...you define truth as compression of contradiction.*"

The amphitheater paused again.

Not in confusion.

But in collective intellectual vertigo.

Because now, truth had been described not as a fixed entity, but as:

- A frame of reference (Einstein)
- A survivor of testing (Feynman)
- A layered energy outcome (Sandeep)
- A feminine enigma (Nietzsche)
- A quantum collapse of probability (AI)
- A moral virtue-in-action (Confucius)

All these truths circled the same question—like satellites unsure of their center.

And standing amidst this swirl, Sandeep listened.

He was not alone.

But he was the weaver. The one who would stitch contradictions into coherence.

He stepped forward again. The arena dimmed and pulsed, not from exhaustion, but anticipation.

The idea of truth—once a throne of marble and certainty—Was now a moving target. And in that movement, it had finally become real.

Because the War of Thoughts wasn't just an intellectual collision anymore.

It was a reconstruction.

Truth is an Effect, Not a Premise

Truth doesn't begin the story—it ends it.

Sandeep stepped forward once more.

The arena still vibrated with the resonance of contradiction—the echoes of Plato's idealism, Einstein's relativism, Feynman's pragmatism, Nietzsche's nihilism, and AI's algorithmic precision.

All brilliant. All incomplete.

He paused—not to silence them, but to let the noise settle into signal.

Then he spoke.

Not louder. Not faster. Just clearer.

"Truth," he said, "is not the beginning of the story. It is the result. The consequence. The echo of reality...in balance."

The amphitheater hushed again—not from command, but from recognition.

"In the framework of Universal Energy Dynamics, truth is not a location in space, not a concept outside time, not a premise that begins inquiry."

He raised a hand. A new hologram emerged at the center of the arena.

Gone was Plato's crisp, Platonic polyhedron.

In its place, a rippling field formed—like the surface of water disturbed by countless droplets.

- Waves collided.
- Patterns emerged.
- Tensions pulsed, and then slowly, from the chaos, a rhythm arose.

A moving harmony—not symmetrical, not fixed—But stable. Cohesive. Alive.

"This," Sandeep said, gesturing, "is truth—not in isolation, but in interaction."

He circled the hologram like a conductor walking around a living symphony.

"Truth is born where energy, perception, intent, and consequence meet. It is not a frozen ideal. It is a living response. It's not what *precedes* reality. It's what remains after reality has unfolded."

He walked the perimeter of the arena now, slowly, deliberately, looking into the eyes of those who shaped human understanding.

- He passed Plato, whose gaze lingered, not in defense but in reflection.
- He passed Einstein, who smiled faintly, already calculating the implications.
- He passed Feynman, who now tapped his fingers—not in impatience, but anticipation.
- Even AI, its spectral form shimmering, began recalculating truth parameters.

"Plato," he said, turning, "You placed truth in the heavens—perfect, untouched, eternal." "Einstein, you saw it bend with velocity and warp with gravity." "Feynman, you watched it disappear until someone dared to measure it."

"But in our world—the lived world— truth is earned. It is not delivered. It is forged."

He paused, letting the holographic field of waves ripple once more.

"Truth is not in the claim. It's in the consequence. It's not in what you believe. It's in what endures after belief has been tested."

As he spoke, something extraordinary happened.

The original Form of Truth—the Platonic polyhedron—still hovered nearby. But now, it began to tremble... not in defeat, but in curiosity.

It drifted toward the UED wavefield.

Then—gently, like light meeting water—it began to merge.

- The perfect geometry of the Form didn't shatter.
- The fluidity of the field didn't collapse.
- They interwove.

Edges curved. Waves smoothed. The fixed met the flexible. The timeless met the temporal. And from that tension, a new form began to emerge:

Not still. Not wild. But responsive.

Truth, redefined—not as premise, but as process.

The amphitheater pulsed—not with applause, but with the recognition of a new era in thought.

Even Plato, who had once seen truth as unchanging, looked up—his eyes reflecting not pride, but revelation.

He reached toward the merged construct.

It pulsed back.

And in that pulse was the answer to a question philosophy had never dared to ask:

"What if truth is not untouchable... but touch-sensitive?"

Sandeep stepped back—not in triumph, but in quiet reverence.

For he had not destroyed the old idea.

He had updated it.

He had shown that truth need not be fixed to be real—That in its capacity to evolve, it had finally become undeniable.

And in that moment, the War of Thoughts had done what no war ever promises:

It had created something new.

The Shift (Not Surrender)

Some truths aren't abandoned. They evolve.

The amphitheater stood suspended—not in silence, but in awe. The air itself seemed to listen. At the center, the merged construct of Plato's eternal Form and Sandeep's dynamic field still floated—no longer a contradiction, but a living harmony.

It pulsed softly—like a heart that had just remembered how to beat.

No longer cold and symmetrical.

Now fluid, aware, and alive.

Plato did not move.

He stood at the edge of his own idea, watching it breathe.

This was not a collapse. It was an awakening.

The philosopher who had once declared:

"Perception is illusion. Only unchanging ideals can be trusted..."

...was now standing in front of a truth that *moved*. And yet, it didn't undermine his wisdom.

It deepened it.

He looked at Sandeep.

Not with resistance.

Not with pride.

But with the humble interest of a mind that had lived too long without being challenged this honestly.

The lines on his face softened—centuries of tension exhaling at once.

Then, quietly—without ceremony, without ego—he asked:

"Then what, Sandeep...is the purpose of knowing truth...if it keeps changing?"

There was no sarcasm in his tone. No challenge. Just genuine curiosity.

A student's question... spoken from the mouth of history's most revered teacher.

Sandeep smiled.

Not with triumph. But with something warmer, deeper—the kind of smile that's reserved for the moment when minds meet without walls.

"To stay aligned," he replied.

He stepped closer, the arena glowing subtly as he spoke.

"To act in harmony with evolving equilibrium. Truth isn't the anchor that holds us down. It's the pivot that lets us move with grace. It's not a commandment. It's a compass. Not a restraint—but a guidepost."

He turned toward the new hybrid model floating above them—a shape no longer bound by lines, but not lost in fluidity either. It spun, shimmered, and breathed with an intelligent rhythm—a geometry of motion, a form of balance.

"Truth may not be permanent," Sandeep said, "but it is never random. Like energy, it follows patterns. It adjusts. It reveals more to those who stay awake. And it always returns to balance, just like the universe."

Plato looked again at the merged truth.

And then—very slowly—he reached out.

This time, the construct did not resist. It responded.

The light softened, then arched gently toward his hand. Not as an object to be held—But as a presence to be felt.

It flickered with soft intelligent light—not glowing with perfection, but with purpose.

Plato closed his eyes.

He wasn't surrendering. He was... remembering.

A truth he had sensed but never named. A cave that hadn't imprisoned—but prepared.

He turned back to Sandeep. And slowly... he nodded.

Not in defeat. Not in concession.

But in recognition.

Recognition of a truth that wasn't lesser than his—but larger. Not fixed in form, but rooted in flow. Not worshipped, but walked with.

At that moment, the amphitheater changed again—subtly.

The cave—so long a symbol of ignorance—didn't collapse.

It opened.

Its walls no longer separated light from shadow.

They channeled them.

And for the first time in history, the shadows weren't deceiving.

They were dancing.

Not to distort.

But to express something real.

The war had not ended.

But its first great battle had found its conclusion.

Not in victory.

In evolution.

Reflection & Transition

The Form had shifted. So had the foundation.

The air within the amphitheater had changed—not emptied, but elevated.

What once echoed with rigid ideals now vibrated with a living rhythm.

At the center, the merged construct of truth floated—not a static form, not a chaotic flux, but a symphony of balance.

It pulsed—gently, steadily—like the heartbeat of understanding itself.

The legends watched in contemplative silence.

Some with narrowed eyes.

Some with folded arms.

Some with open minds.

But none unmoved.

Plato remained standing. Still watching. Still reflecting. He had not been defeated. He had been expanded.

And Sandeep—no longer challenger, but orchestrator—turned slowly, his gaze sweeping across the arena like a comet arcing across a night sky of minds.

"One truth down," he said, voice quiet, but cutting through the amphitheater like a blade through silk. "Many to go."

A ripple spread—not through sound or light, but through awareness.

The kind of ripple that doesn't disturb the air—but stirs the soul.

And then, as if summoned by the next gravitational necessity, the amphitheater began to reshape again.

The stone floor trembled—not violently, but like a clock reconfiguring its rhythm. Geometry melted into calculation. Marble gave way to math.

A new structure emerged—not sculpted by hand, but rendered by reason.

- Its frame composed of spinning numbers.
- Its arms woven from field equations and light cones.
- Its seat: a throne of pure relativity—curved, elegant, pulsing with silent logic.

And seated upon it, already waiting, already smirking...

Albert Einstein.

His hair wilder than usual. His posture relaxed but ready. A floating notebook hovered beside him—its pages filled with equations that rearranged themselves in real time, reacting to the moment like living language.

He tapped his foot gently—perhaps to a rhythm only spacetime could hear.

Then, grinning like a child who knew the punchline before the setup, he leaned back.

"I suppose," he said, "now it's my turn...to fall into the gravity of your thoughts, Sandeep."

Laughter flickered through the chamber—light, real, and warm.

- Feynman leaned forward, like a kid watching his favorite magician take the stage.
- Buddha remained still, but his serenity now carried a sense of quiet readiness.
- Krishna gently rested his fingers on his flute—not to play, but to remind the moment that silence, too, was music.

And somewhere, in a realm unbound by time, gravity itself stirred, sensing its role about to be reconsidered.

Then—unexpectedly, but naturally—another voice rose.

Not loud.

Not demanding.

But *present.*

A voice not of argument, but of wisdom wrapped in timeless grace.

Krishna stepped forward.

Until now, he had watched—smiling, listening, waiting.

But now... the Form had evolved. And so had the moment.

His presence shimmered—not supernatural, but supra-logical.

And as he looked toward the center, his smile deepened—not in amusement, but in recognition.

"Truth," he said softly, "is not what you *know*. It is what your karma reveals."

The words landed gently, yet with seismic resonance.

"It is not born in silence—but in action. Not discovered through detachment—but through intelligent alignment with life."

He turned to Plato, who bowed his head—Not to a god, But to a fellow seer Who had named what he had sensed, but not spoken:

That truth, too, walks the path of karma.

Then Krishna looked at Sandeep, eyes glinting with ancient mischief and cosmic love.

"And you," he said, "have simply reminded the world what the Gita has always whispered:

Yathā karma tathā satyaḥ. 'As is the action, so is the truth.'"

The arena didn't erupt.

It pulsed.

Not with applause.

But with recognition.

Not with noise.

But with a quiet upgrade of consciousness.

Because in that moment, the war of thoughts revealed its true face:

It wasn't a battle against ancient wisdom.

It was the next step of it.

A long-overdue evolution.

Not destruction, but integration.

Not opposition, but alignment.

And in the center, the Form of Truth glowed—still pulsing.

Still responding.

Still becoming.

Chapter 2: Gravity – Force or Fiction?

Hot Seat: Einstein

Where falling means more than physics.

The truth had pulsed.

The Form had bent.

The philosopher had nodded.

But the amphitheater did not rest. It never does.

Even as the final vibrations of Krishna's voice faded—

"Yathā karma tathā satyaḥ"— another pulse surged through the chamber.

But this one was different.

It didn't echo like a word.

It didn't shimmer like a thought.

It thudded—slow, heavy, deep—like the heartbeat of the cosmos itself.

The arena began to shift.

Where moments ago the floor had mirrored the geometry of ancient Greece, it now reconfigured into curvature.

Smooth stone gave way to a surface that *bent* rather than *broke*.

Field equations etched themselves across the walls in glowing arcs.

Vectors twisted into spirals.

Light itself bent—subtly, but undeniably—toward a new center.

The boundaries of the chamber dissolved, giving way to an impossible observatory—Open space stretched beyond the amphitheater walls.

Above them, stars swirled like probabilities, and in the distance, black holes rotated in solemn majesty—silent, devouring, unblinking.

Gravity had entered the arena.

And at the center, already seated—as if he had always been there—was Albert Einstein.

His wild hair looked even more rebellious under the starlight.

His eyes sparkled—not with arrogance, but with anticipation. A hovering notebook orbited beside him, filled with field equations that rewrote themselves, reacting to the energy in the room like particles to a field.

His seat was not a throne. It was a construct of spacetime itself—elegant, curved, and *impossible to diagram* in Euclidean terms.

He looked at Sandeep. And smiled.

"I suppose," he said, with that legendary glint in his eye, "now it's my turn to fall...into the gravity of your thoughts."

Laughter rippled through the crowd—Not light. Not forced. But nervous. Expectant.

Because everyone knew: This wasn't just the next debate. This was the confrontation of a pillar—one of the most sacred and stubborn truths of modern science.

From the shadowed edge of the arena, another figure emerged.

Isaac Newton.

Silent. Composed. The apple still in his hand. But this time, it didn't symbolize inspiration. It symbolized legacy—and perhaps... the weight of tradition.

He stood still, watching.

Waiting.

Sandeep stepped forward, eyes fixed not on the apple, but on Einstein.

Not with reverence.

But with readiness.

"Let's talk about gravity," he said, voice steady, "Not as a law. Not as a curve. But as a question—One that has pulled us down for far too long."

Einstein raised an eyebrow. Newton gripped the apple just a little tighter.

The amphitheater shifted again—light bent inward. Equations swirled around them like gravitational lenses.

The second battle in the War of Thoughts had begun. Not against mysticism. But against the most elegant illusion ever conceived in the name of certainty:

Gravity.

The Chair of Curved Space

Where even space bows to mass, and time forgets itself.

The War of Thoughts was shifting. Not just in tone, but in terrain.

From timeless Forms and dynamic truths...To the very fabric of existence—To gravity, where even space bends, and time forgets who it is.

The amphitheater began to transform once more. But this time, it didn't rebuild. It folded.

Straight lines curled like they'd grown tired of pretending. Angles dissolved into arcs. The stone beneath the audience's feet rippled softly, as though it were cloth stretched across invisible mass.

Above them, the ceiling unfurled into a vast, velvet-black canvas—But it wasn't empty.

It was dense—with meaning, with memory, with math.

Celestial equations floated overhead like constellations—Einstein's field equations looping and spiraling through the void, their symbols glowing like ancient sigils from a scientific scripture.

- Tensor notations blinked in and out of view—stitched across space like cosmic graffiti.
- Some pulsed with quiet confidence.
- Others spiraled, mimicking the graceful fall of dying stars.

At the edges of the arena, light began to slow. Time stretched—not linearly, but like taffy under pressure. Clocks melted. Trajectories bent.

Reality had lost its rigidity. It had become... relativistic.

In the distance—though "distance" now felt like a soft suggestion rather than a measurable unit—black holes hovered, silent and majestic. They didn't roar. They didn't pull. They simply existed—with the terrifying authority of something that needed no permission to devour light.

One of them pulsed faintly—Not as a threat, but as if listening.

And at the very center of this curvature, sat the man who bent the universe without lifting a hand.

Albert Einstein.

Perfectly relaxed. One leg crossed. Fingers steepled together, as though calculating the curvature of the moment itself.

His chair was not made of wood or stone. It was a geometric impossibility—a seat shaped from the very distortion of spacetime.

- Its arms shimmered with gravitational tensors.
- Its back arched like a time dilation curve caught mid-warp.
- And beside him floated a notebook—its pages turning themselves, formulae reshaping as if they were alive and listening.

Einstein didn't sit like a man prepared to defend a theory. He sat like a man who had already dropped the apple...and was waiting for the universe to explain itself.

The atmosphere had changed.

This was no longer a poetic exchange of ideals.

This was physics.

And the minds gathered felt it.

- Feynman leaned forward, practically buzzing like a particle in excited state.

- Hawking's avatar flickered sharper, his attention fully locked in.
- Newton, now fully visible, twirled the apple in his palm—his face unreadable, somewhere between nostalgia and suspicion.
- Krishna stood silently at the edge, flute resting against his shoulder—not playing, just observing.
- Even Plato, still reverberating from the collapse of his static ideal, leaned in—not to teach... but to learn.

Sandeep stepped forward.

He passed through the field of floating symbols, the broken clocks, the light rays curling like tendrils of thought.

And he stopped in front of Einstein—Where certainty had once lived... and now waited to be questioned.

"We've bent the truth," he said softly. "Let's see what happens... when we bend the space around it."

Einstein smiled.

The same smile he had worn when he first imagined riding on a beam of light.

His chair—his throne of curved reason—tightened subtly beneath him, as if preparing itself for the impact of what was to come.

The next ripple in the War of Thoughts had been summoned. And its name was gravity.

Gravity as Geometry

What if falling isn't being pulled—but simply following the shape of reality?

Einstein leaned forward, his fingers brushing away an invisible speck from the edge of his floating notebook, though nothing ever landed on it. The equations swirling around him formed a soft halo—numbers, tensors, and curvature dancing like light particles around a star.

He smiled. Not with arrogance. Not with certainty. But with the quiet joy of a teacher...about to show the world how reality bends.

"Everyone," he began, his voice warm and rhythmically precise, "talks about gravity as a *force*. But that's only because we've spent our whole lives being pulled down." "What if I told you... that nothing pulls you at all?"

He lifted his hand.

The amphitheater dimmed, like a planet waiting for its sun to rise in math.

A visual model unfolded at the center—A massive grid suspended in three-dimensional space. Elegant. Meshed. Infinite.

And at its center, a glowing sphere of mass settled gently onto the grid—Sinking. Bending. Curving the fabric around it like a bowling ball dropped onto a stretched membrane.

"This," he said, gesturing, "is spacetime. And this..."—he pointed to the gentle depression—"is what mass does to it."

Another smaller object appeared near the edge of the grid—satellite, moon, apple, it didn't matter.

It began to move—Not by force, But by following the curve.

"There is no invisible hand dragging this object inward. It simply moves along the straightest path available—A geodesic, dictated by the shape of reality."

The audience leaned in.

Even light bent subtly near the indentation, arcing with humility.

Einstein raised his hand again, and with a snap, a beam of light shot through the grid. It curved—not wildly, but just enough to betray the lie of flatness.

"Even light," he grinned, "can't escape—not because it's held back, but because the space it moves through...is not flat anymore."

The model dissolved. Replaced by a black-and-white holographic scene—1919. Total solar eclipse. Starlight curving around the sun. The first visual proof of a theory that had been declared insane.

The arena remained still. Reverent.

Then came a ripple—soft at first, then rhythmic. Like the surface of a cosmic pond disturbed by a celestial quake.

"And then," Einstein said, "came the gravitational waves. Ripples in spacetime. Caused by black holes colliding—Detected a hundred years after I predicted them."

He leaned back now, his arms resting effortlessly on the curved arms of his spacetime chair, like a conductor who had just finished his first movement.

"So, my dear Sandeep," he said with a twinkle, "gravity is not a force that pulls you to Earth. It's the Earth telling spacetime how to bend—and spacetime telling you how to move."

A few nods rippled through the crowd.

- Feynman clapped once—thoughtfully, not theatrically.
- Hawking's avatar pulsed, signaling respect.

- Newton twirled the apple, unfazed but clearly unconvinced.

Einstein looked at Sandeep now, not as a victor, but as someone who had just shared the architecture of his soul.

"It's not magic," he said with that timeless grin. "It's geometry."

The amphitheater paused. Basking in the elegance. In the coherence. In the clean curve of understanding.

But Sandeep was already stepping forward.

Because elegance is not enough—Not when the very geometry breaks.

And it does.

Not everywhere. Not always. But in the one place where all lines vanish, all rules falter, and all certainty collapses:

Inside a black hole.

The Paradox of the Black Hole

What bends when the thing that bends... no longer exists?

Sandeep didn't answer right away. He circled the floating model of curved spacetime, his steps silent, deliberate. Objects still spiraled inward. Light bent gracefully. It was beautiful—almost too beautiful.

His gaze was focused, contemplative. But behind his eyes, something sharpened. Not to cut, but to pierce the edge of understanding.

Then he stopped. And turned to Einstein.

"If gravity is curvature," he asked, voice steady, "what happens when spacetime itself breaks down?"

The question didn't echo. It landed—with a thud.

Not in the mind.

But in the gut.

Einstein's smile faded, replaced by the quiet stillness of calculation. Newton leaned forward. Hawking's avatar flickered with heightened frequency, as if his digital form itself reacted to the question. Even Krishna tilted his head slightly, as though intrigued by how this thread of karma would unfold.

Sandeep stepped fully into the center of the arena.

With a wave of his hand, Einstein's model froze—the elegant curvature suspended in mid-demonstration.

Then, slowly, a new simulation began to rise.

Darker. Denser. Inevitable.

A massive star appeared—spinning, seething. Its surface shivered as internal pressure collapsed. And then—implosion.

The star folded in on itself. Its mass compressed. Its light curled inward, spiraling like smoke into a cosmic drain.

"This is not a thought experiment," Sandeep said. "This is the most real, most extreme thing the universe offers us: a black hole."

The audience watched in hushed awe.

- The singularity formed.
- Light ceased escaping.
- Time distorted beyond recognition.

“Outside the event horizon,” Sandeep continued, “spacetime bends. Light curves. Time slows.”

“But inside...” He paused.

“Inside, everything inverts.”

The simulation zoomed in—past the horizon, into the unspeakable.

The once-smooth grid of spacetime began to fracture.

- Axes spun like unhinged compasses.
- Metrics twisted into paradoxes.
- Geometry, once elegant, became meaningless.

“Time becomes space. Space becomes time. Coordinates collapse. And curvature...”

He turned to Einstein.

“...has no meaning anymore.”

The crowd didn’t move.

Sandeep pressed forward—not with defiance, but with precision.

“How can gravity be curvature,” he asked, “when the very canvas required to express it has been destroyed?”

“What bends...when there’s no ‘direction’ left to bend in?”

He paused. Einstein looked on—not threatened, but... aware.

“General Relativity works—beautifully—up to the edge. But beyond the event horizon, it doesn’t just break...it vanishes.”

The simulation faded into darkness.

But the black hole remained.

It pulsed in the background like an unsolved riddle. Not loud. Not violent. Just... present.

"And yet," Sandeep said, "mass is still there. Energy is still there. Something real is still happening."

He turned to the audience.

"But it's not curving anything. Because curvature needs structure. It needs space. It needs time. And a black hole devours both."

A hush fell. Feynman's coin stopped spinning. Turing murmured something about computational collapse. Plato leaned forward—curious. Buddha watched with stillness that understood paradox.

Sandeep let the silence settle like dust after an explosion.

Then he concluded—not with drama, but with finality:

"Inside a black hole, gravity doesn't intensify." "It ceases to exist as we've defined it." "Not because mass is gone—but because spacetime—the medium through which curvature expresses itself—has been annihilated."

The amphitheater sat frozen in thought.

Einstein said nothing.

Not from defeat.

But because he knew—this was the place even his equations refused to go.

This was the limit. The horizon. The line between elegance and unknown.

The geometry had cracked. The model had failed. And now—it was time to rebuild. Or redefine.

The Original Force Returns

Some ideas fall with apples. Others fall apart in singularities.

As the echoes of Sandeep's final words dissolved into the curved silence of spacetime, the amphitheater shifted again. But this time, the change was not warped or bending. It was measured.

Precise.

As if the universe, after spiraling into abstraction, was now calling for grounding.

A tremor passed beneath the arena floor. Not chaos. But order—the return of symmetry, of law, of clean predictability.

From the edge of the stage, a new seat rose.

Not floating. Not fluid.

But angular, definite, carved from stone with razor-sharp clarity. Its edges bore chiseled numerals, Latin inscriptions, and geometric diagrams. It cast a shadow—not rippling like Einstein's, but pointing true.

A man stepped into the light. Tall. Stern. Composed. His coat crisp, his posture regal. Isaac Newton had returned—not as a relic, but as a force of nature.

In one hand, he held a scroll—the *Principia Mathematica*, its golden equations glimmering as if freshly inked by the cosmos itself. In the other, effortlessly casual—An apple.

He looked at Einstein, and smirked. Then at Sandeep, with curiosity.

The amphitheater held its breath.

Then, in a voice as exact as a pendulum, Newton spoke:

"Gravity," he said, "is not a distortion. It is a force. Real. Measurable. Proportional to the product of two masses, and inversely proportional to the square of the distance between them."

He tossed the apple into the air. It rose. It fell.

"Simple. Universal. Irrefutable."

A murmur passed through the audience. Confucius nodded—appreciating the elegance. Aristotle smiled, as if hearing his own mind echoed centuries forward. Even AI flickered momentarily, reindexing Newton's constants.

Einstein raised an eyebrow, amused.

"Isaac," he said softly, "you knew more than your time allowed. But mass doesn't pull—it shapes."

"And yet," Newton replied, "the apple still falls. Not because it follows geometry—but because it is acted upon."

He held the apple up.

"Action at a distance, if you must. But action, all the same."

The room was divided not by sides, but by centuries of interpretation.

Sandeep stepped forward. Not between them—among them.

"And what of the black hole, Newton?" he asked. "There's mass—but no direction. No field. No line. No pull. What acts when there's nowhere for interaction?"

Newton didn't flinch.

"Then the force is not absent," he said. "It is concentrated. Too strong to escape—not too weak to exist."

Einstein interjected.

"Concentrated where? At a point with no dimension? Without space, there is no direction for force to act. You are describing tension in a void."

The arena stirred again. The audience wasn't watching a debate. They were witnessing a collision of models.

Sandeep raised his hand.

A new hologram formed—a triangle. Each corner labeled:

- Force
- Curvature
- Collapse

At the center: Energy in Equilibrium.

"You're both right," he said. "And both incomplete."

He walked to the center, where the triangle hovered above the floor.

"Gravity is not a force, nor is it geometry. It is the consequence—of energy seeking equilibrium through space."

He turned to Newton.

"When space exists—yes, mass interacts. And we experience it as force."

He turned to Einstein.

"When space curves—yes, objects move along that curve. And we experience it as geometry."

Then he pointed at the third corner—Collapse.

"But inside a black hole—when space itself is gone—there is no force to pull, no shape to follow. Only the residue of energy unable to express itself."

A deep silence followed. Not confusion. Not surrender. But integration.

The three men now stood at the triangle's corners—Newton. Einstein. Sandeep.

Each with their truth. Each with their limits.

And the war of thoughts... had changed again. From debate to convergence. From collision to reconstruction.

Because when all models crack at the edges...

It's not the universe that must change. It's the framework we use to explain it.

And that framework...was about to be rewritten.

The Black Hole Simulation – Thought Experiment

Let the stars collapse. Let the light vanish. Let truth be tested in the singularity.

Sandeep raised his hand.

The arena darkened. Not just the lights—but the equations themselves dimmed, as if mathematics, too, was holding its breath for what was about to be shown.

This was not philosophy. Not allegory. Not curvature. This was the collapse of meaning under extreme reality.

A new simulation unfolded at the center—Not abstract. Not symbolic. But visceral. Elemental. Real.

A massive blue-white star appeared—pulsing with radiant life. It glowed with the dignity of fusion, the symmetry of balance, the confidence of a celestial engine that had burned for eons.

Sandeep spoke softly.

"Let's take one of the universe's most honest expressions...A dying star."

The simulation zoomed closer.

- Fusion slowed.
- Core pressure dropped.
- The shell trembled.

Then—collapse.

Not explosive. But inevitable.

The core buckled. The layers folded inward. Light bent violently, drawn toward the forming pit.

Through the center of the star, a glowing ribbon of time had been threaded—visible now as a golden thread unraveling, stretching like a pulled memory.

Sandeep's voice guided them—not as a lecturer, but as a witness.

"Gravity, they say, is strongest here. Then why..." He paused.

"...why does light not escape?"

The room held its breath.

He turned toward Newton.

"Shouldn't the force push harder? Like a spring compressed to infinity?"

Then to Einstein.

"Shouldn't the curvature snap back? Or at least ripple violently from the pressure?"

But what they saw... was different.

- Light didn't burst outward.
- Time didn't scream.
- They both faded.

Not shattered. Not destroyed. Just... gone.

Sandeep stepped closer to the center of the arena.

"Or perhaps," he said gently, "it's not gravity pulling light in—but spacetime itself dissolving around it."

He turned to the audience.

His voice sharpened, not in tone, but in clarity.

"You see, we've assumed gravity is a cause. A starter of motion. An invisible hand pulling things toward each other."

"But in the framework of Universal Energy Dynamics, that assumption is reversed."

He gestured again.

The simulation zoomed inward—past the event horizon.

No light. No time. No spin. No curve.

Just stillness.

A terrifying, dignified, *necessary* stillness.

"Inside a black hole," Sandeep continued, "there is no force pulling. No curvature to bend. No geometry to follow.

No field to align with."

"There is only collapse—A state in which energy no longer requires a medium to interact."

"What we called gravity has fulfilled its role—Guiding imbalance into stillness."

He paused.

Then:

"The need for gravity is over. It doesn't explode. It doesn't fight. It simply... ceases."

A new model appeared now, overlapping the old one.

This wasn't spacetime. It wasn't force. It wasn't curvature.

It pulsed—like breath. Like thought. Like resolution.

UED in motion.

A radiant diagram appeared, pulsing softly with transitions of states:

- Imbalance → Motion

- Motion → Curvature (when space is available)
- Motion → Force (when resistance is present)
- Equilibrium Reached → Gravity Vanishes

"Gravity," Sandeep said, "is not a thing. It's not a law. It's not a pull."

"It is the resistance of energy on its path to equilibrium."

"When that balance is achieved—especially in extreme environments like black holes—gravity doesn't intensify. It evaporates."

The screen faded.

And what remained was the black hole—not as a monster, But as a return point. A cosmic hush. A memory folded into silence.

No scream. No shadow.

Just... peace.

And in that silence, a whisper moved through the amphitheater. It wasn't spoken. It was felt.

What if gravity was never what we thought it was?

The war of thoughts didn't roar.

It recalibrated.

Because when the pull is gone, And the curve is broken, What remains is not emptiness—It is the source waiting to be understood.

The Breakdown – Einstein Concedes a Gap

When the fabric no longer stretches, even the tailor must pause.

The simulation faded.

The swirling collapse of the black hole, the bending of light, the trembling grid of spacetime—all dissolved into silence.

The amphitheater dimmed, returning to a low, contemplative pulse. No longer an arena—now a cathedral of thought. And at its center, only the soft echo of Sandeep's final words remained:

"Gravity is the resistance of energy on its journey to equilibrium."

The words didn't just linger. They settled—into minds, into frameworks, into the places no equation had ever touched.

All eyes turned to Einstein.

He didn't speak immediately.

He sat still—unmoving, unreadable. The usual spark in his eyes, the signature grin playing at the corners of his mouth...Gone.

In its place was something far rarer. Something sacred.

Reflection.

Not the silence of being proven wrong—but the silence of being shown a truth you didn't know you were waiting for.

His chair of field equations now wrapped around him differently—no longer a shield of logic, but a memory of movement.

Einstein straightened, the weight of unspoken thoughts coalescing behind his eyes.

Then, he looked toward the center of the arena—at the place where the black hole had once been. Where the laws of physics had collapsed into quiet. Where understanding had stopped... and awareness had begun.

He spoke.

"Relativity," he said, slowly and with solemn clarity, "does not describe what lies beyond the event horizon."

The words did not fall like surrender. They landed like truth. Heavy. Simple. Clear.

A scientist's highest honor: Intellectual honesty.

He leaned back in his seat, the equations on his notebook now gently dimming—like stars turning inward.

"I gave the world a geometry that could bend, stretch, ripple..."

"But in the singularity..." "There is no geometry. There are no metrics. No frames of reference. Only conditions beyond understanding."

He paused. The pause was not hesitation. It was reverence.

"I bent space and time," he continued. "But I never imagined they could disappear."

Then, quietly—almost to himself, but loud enough for the cosmos to hear:

"I always knew general relativity was incomplete...I just never thought I'd hear someone explain why...so simply."

The amphitheater remained still.

Not in shock. Not in celebration. But in something far rarer: Unity.

Behind him, Plato gave a subtle nod—a gesture only a philosopher could truly understand.

He had seen it before: When an idea—proud, powerful, and poetic—meets its boundary.

Not in disgrace. But in grace.

Not in ruin. But in relevance.

From his corner, Feynman muttered,

"Well, damn."

Newton looked away.

Not dismissive. Just... quiet. The kind of quiet that happens when certainty is no longer certain.

And yet, the moment was not defeat.

It was recalibration.

Einstein turned to Sandeep.

There was no defensiveness in his gaze. Only gratitude.

"Maybe gravity is not the law that governs the universe," he said, "but a consequence...of what the universe is trying to balance."

And in that sentence, a pillar of modern physics transformed—From doctrine to dialogue. From authority to evolution.

For a moment, the arena felt...

Humble.

As if even the stars above had paused. Not to shine. But to listen.

The Core Argument – Gravity is Fictional Friction

Not a pull. Not a curve. Just resistance... along the path to stillness.

Sandeep stepped forward once more.

Not triumphant. Not aggressive. Just steady.

The kind of calm that no longer seeks to win an argument, but to illuminate the edges of shared ignorance.

Around him, the amphitheater had settled into something rare—not silence, but readiness.

- Newton stood quiet—skeptical, but listening.
- Einstein, still seated in his chair of curved spacetime, watched without defense.
- Plato, hands folded, leaned slightly forward, as if truth had taken a form he hadn't seen before—but recognized.

Even the floor beneath Sandeep's feet responded. It stabilized, just slightly—like the amphitheater itself was finding its own center of mass.

He exhaled, then spoke.

"Let's call gravity what it really is."

"Not a fundamental force. Not even a curvature of space. But a perceived effect—A byproduct of energy being redirected by imbalance."

He raised his hand.

A new model materialized—minimalist, elegant, unassuming. No spinning galaxies. No falling apples. Just a single wave of energy.

It pulsed across the open space—fluid, pure, intentional. As it moved, it encountered interruptions:

- Matter
- Resistance
- Asymmetry

With each encounter, the wave distorted—curved, deflected, slowed.

Observers around it—tiny silhouettes of minds—began labeling the patterns:

- They called the deflection a *pull.*
- They wrote equations.
- They named it: Gravity.

Then the simulation zoomed in.

The medium was removed.

All surrounding interference—gone. The wave moved again—through a uniform field. No bending. No resistance. Just pure motion in perfect equilibrium.

"You don't see gravity here," Sandeep explained, "because nothing interrupts the energy's intent."

"There is no imbalance to correct. No distortion to compensate. No tension to resolve."

He turned to the amphitheater—now still, absorbing every word like a sponge made of stars.

"Gravity only appears where equilibrium is disturbed. Where mass exists—not as a source of pull, but as a site of energetic asymmetry."

"Energy does not move because it's pulled. It moves because it is trying to restore order."

A new projection unfolded beside the wave model.

A black hole—collapsed, silent, beyond field lines.

No spirals. No vector arrows. No curvature grids.

Just... rest.

"Inside a black hole," Sandeep continued, "there is no medium left to express curvature. No reference for resistance. No coordinates to define direction."

"What remains...is not force. Not geometry. But the final stillness of energy."

"At rest. Or in collapse. It makes no difference. Because either way, gravity has fulfilled its role."

He paused.

Let the words breathe.

"We mistake that stillness for infinite gravity. But in truth...it is the absence of reaction."

"It is not strength. It is completion."

From the corner, Feynman whispered,

"So it's not that gravity is infinite...It's that it's done."

Einstein nodded—subtly. Not with concession. But with the kind of respect reserved for clarity he hadn't yet reached on his own.

Newton remained still, holding his apple like a relic of belief—a model that once reshaped the world, but now felt too solid for this new, fluid truth.

Sandeep stepped back slightly.

Not to retreat. But to complete the idea.

"Gravity is not a thing. It's not a force line. It's not a curvature in need of measurement."

"It is a trace. A kind of fictional friction—a temporary resistance where energy still hasn't found peace."

He gestured gently toward the quiet black hole projection.

"And once it does—once balance is reached—gravity doesn't roar. It doesn't collapse. It simply... disappears."

"Not because it was defeated—but because its purpose was fulfilled."

The amphitheater pulsed.

Not from technology. Not from light.

But from the collective resonance of realization.

Some in the crowd shivered. Not from fear.

But from the overwhelming feeling that they had just understood something ancient...for the very first time.

The Independent Equation

When recognition comes from the mind that once bent space itself.

Einstein leaned forward again. But this time, it wasn't to object. It wasn't to challenge.

It was to recognize.

His notebook, once alive with equations and shifting tensors, now floated closed beside him—like a well-used violin finally allowed to rest.

And his eyes—eyes that had once glimpsed the structure of spacetime itself—now softened.

No longer the relativist. Now, the elder.

He looked directly at Sandeep. And for the first time in this long, monumental debate, his voice carried not the weight of scientific elegance, but the warmth of personal admiration.

"My dear," he began, "I've gone through your report—your work on black hole energy dynamics, Chavan's Law, the so-called Chavan's Limit."

The amphitheater stirred.

Even Newton looked up—curiosity overriding reservation at the mention of new laws.

Sandeep stood motionless.

Einstein continued.

"You are... an independent researcher. Just like I was."

"No ivory tower. No polished faculty hallways. No peer-reviewed gatekeeping."

"Just ideas—naked, bold, and stubborn."

He smiled now, the kind of smile reserved for those who know the loneliness of discovery.

"Let me tell you from experience...it will take time. Be patient. Even truth takes time to orbit belief."

A hush swept through the chamber.

Even among legends, the weight of such acknowledgment was felt.

But then... came the twinkle.

That familiar glint in Einstein's eye. The one that meant: *Here comes the fun part.*

He sat up straighter, grin spreading across his face, and leaned forward—not with resistance, but with playful revenge.

"But tell me, Sandeep..."

"If gravity dissolves into energetic stillness beyond the horizon..."

He raised an eyebrow.

"Then where—the hell—does the light go?"

Laughter flickered across the room—not derision, but delight. The spirit of Socratic mischief had re-entered the stage.

Einstein's finger lifted in the air like a conductor's baton.

"If energy doesn't vanish...And there's no space to curve, No force to act—Then what holds it? What contains it? What decides its fate?"

He pressed further, grin widening.

"And if equilibrium is the destination...Then who—or what—set the coordinates?"

The challenge wasn't to destroy. It was to deepen. To open the next layer.

Sandeep smiled, fully ready to respond.

But before he could speak, a sound—low, gentle, resonant—vibrated through the amphitheater.

Not a sound from instruments. A sound from within. Like a flute played through the soul of space.

Krishna stepped forward.

He shimmered—not in light, but in understanding.

He didn't walk like a god. He walked like someone who had watched stars burn, collapse, and become—and still chose to smile.

He faced Einstein.

"The light," he said softly, "does not go anywhere, Albert. It returns. It folds. Not into disappearance...but into memory."

The room held its breath—not from confusion, but from a sense that something ancient was being revealed.

Krishna stepped into the center, where the last projection of the black hole still hovered—silent, still, misunderstood.

"A black hole is not a prison," he said. "It is a passage. A purifier."

"The universe does not allow energy to vanish. It transforms it—in silence."

"What you see as collapse...is actually containment with purpose."

The visual behind him responded.

The black hole's perimeter shimmered—not as an event horizon, but as a boundary of transition.

Krishna turned now, slowly, toward Sandeep.

"And the writer of that transformation..." "Is not a god...But equilibrium itself—Guided by the laws your friend here has begun to touch."

He glanced around the amphitheater.

"Black holes are not the end of gravity. They are where gravity bows out—And intent takes over."

A pulse moved through the room.

Not applause. Something deeper.

A knowing. A recognition.

Sandeep stood still—not because he lacked an answer, but because Krishna had answered for the universe itself.

Einstein sat back. His hands folded. His grin softened.

"Alright," he said quietly, "Now I see... Where the equations end...and the vision begins."

Above them, the stars seemed to blink slower. Not out of exhaustion—But reverence.

The amphitheater was no longer just a stage.

It had become a threshold.

And now...

It was ready for the conclusion.

The Gauntlet – When Equations Demand Answers

Krishna's words had lingered like music in the void—resonant, deep, complete.

But science does not end with stillness. It responds to stillness with questions.

Just as the amphitheater seemed to absorb the calm of Krishna's revelation, a flicker of static broke the silence.

It wasn't a glitch. It was a summoning.

The space near Einstein rippled—A ripple unlike any seen before. Not curvature. Not collapse. Computation. Compression. Quantum entropy made visible.

The ripple solidified into form.

Stephen Hawking had arrived.

His avatar wasn't a person. It was a paradox wrapped in photons—a shape built of distortion and brilliance. His presence didn't bend the room. It interrogated it.

And then came the voice.

Not warm. Not melodic. But clear. Commanding. Like a man who had waited his entire existence to finally ask what others were too afraid to.

"Sandeep," the voice echoed, "Krishna speaks in intent. But I speak in information."

A pause.

"What exactly happens inside a black hole—in your theory?"

"If not gravity, then what causes the collapse?" "Where does the energy go?" "Is it conserved? Does it radiate? Does it encode?" "Or is this just philosophical sugar over scientific uncertainty?"

The arena shifted again—not in shape, but in atmosphere.

The weight of precision had returned.

Hawking wasn't here for reverence. He was here for the math.

Einstein leaned in again, this time electrified.

"And I'm still not done," he said, now energized. "If space collapses...How do your energy states transition?" "Where is your reference frame?" "Where does entropy go?" "What happens to causality when spacetime unravels?" "What—exactly—is the condition beyond the event horizon?"

The questions fell like meteors. Not destructive. But dense. Heavy with the mass of intellectual urgency.

This was no longer philosophical sparring. This was the gauntlet.

The arena turned again—this time fully—to Sandeep.

The independent researcher. The challenger of frameworks. The architect of UED. Now standing at the eye of the intellectual storm.

Around him:

- Einstein, brilliant and hungry.
- Hawking, sharp and relentless.
- Newton, quiet but observant.
- Plato, patient.
- Feynman, bouncing mentally across probabilities.
- Krishna, silent now—his part spoken, his gaze centered.
- And all of reality's paradoxes... circling like question marks with weight.

They no longer wanted philosophy.

They wanted pattern. They wanted process. They wanted proof of principle.

"The universe has collapsed," their eyes seemed to say. "Now show us how it continues."

This wasn't a test of memory.

It was a test of origin.

Of energy without space, of information without time, of motion without force.

And Sandeep—unarmed but unfazed—stood in the middle of it all. Not defending an idea. Becoming it.

Because now, it was time for UED to speak.

Not as theory. But as the last language left when all others fail.

Sandeep Responds – The Collapse of Form, The Rise of Understanding

When curiosity turns into cross-examination, only clarity can answer.

Sandeep stood motionless at the center of the amphitheater.

Not in hesitation—But in listening.

The questions had come not as challenges, but as reverberations of necessity. Each one—Hawking's, Einstein's—had carried the weight of unsolved paradox.

Now, the air around him stilled—Not from silence, but from expectation.

And then, Sandeep stepped forward.

Not to dismantle a theory. But to construct an understanding.

"Thank you, Stephen. And Albert. You've asked what must be asked. And now, let me answer—Not from myth. Not from metaphor. But from the lens of Universal Energy Dynamics."

Behind him, a new model formed.

It wasn't curved. It wasn't torn. It didn't spiral or collapse.

It was layered—a stratified field of energy states, each rotating in harmonic proportion. A cosmic gyroscope—balanced, breathing, alive.

"Let's start at the source: *How are black holes formed?*"

A star appeared—burning bright, pulsing outward in equilibrium. A living system balancing internal pressure with outward radiation.

"A black hole forms when that equilibrium is no longer sustainable. When the internal energy of a system exceeds its capacity to maintain spatial balance, the result is not a pull, but a surrender—a collapse inward, not caused by gravity, but by the loss of resistance to compression."

The simulation unfolded:

- The star's pressure faded.
- Radiation weakened.
- Energy folded inward in layers—not sucked, but withdrawn from expression.

"In UED, gravity does not *pull* this mass inward. Rather, resistance vanishes. And when there's nothing left to oppose collapse, energy condenses into its most compressed, inert state."

The model rotated. Another layer unfolded.

"Now—*what happens inside a black hole?*"

The model zoomed inward, through the event horizon.

"Inside, energy is no longer expressed through space or time. It is not destroyed. It is inactive—held in what I call the Energy Memory Field (EMF)."

A glowing shell formed around a dark center—still, silent, preserved.

"This is not entropy. It is not chaos. It is energy... waiting."

Above the model, a term lit up:

Chavan's Law:

Energy cannot express force unless spatial imbalance exists. In the absence of space, force collapses into potential.

Then came another projection: a red boundary line labeled:

Chavan's Limit

The point beyond which energy transitions from expressive to potential mode due to spatial breakdown.

Hawking leaned forward.

"And the light?" he asked.

Sandeep turned to him.

"Light does not go anywhere. It does not vanish. It detaches from reference."

"No space. No path. No observer. It is preserved—not in motion, but in state. Temporarily unreachable, but not lost."

Another question flashed: *Is information lost?*

Sandeep nodded respectfully toward Hawking.

"Your question, Stephen, about information, is essential—But it rests on the assumption that information must always remain expressed."

"In UED, information is never destroyed. It is compressed—into a non-expressive energy state. Not entropy. Not radiation. But what I call inert intelligence."

A pause.

"Like a paused thought. Like a word held at the edge of the universe's tongue."

The amphitheater was utterly still.

No mind wandered. No eyes blinked.

Sandeep continued.

"Gravity, then, is not the cause of collapse. Collapse causes gravity to appear—as friction in transition. Once the transition is complete, that friction ends. Gravity 'disappears'—not because it failed, but because its task is fulfilled."

A new simulation formed: A black hole. Still. Smooth. No vectors. No warping lines. Just peace.

"A black hole is not a mystery. It is a conclusion. The universe's way of restoring silence after imbalance."

The model faded.

The arena dimmed.

The silence that followed wasn't empty.

It was honest.

Hawking's avatar blinked slower—processing, absorbing, reevaluating.

Einstein leaned back.

His smile returned—softer now, but more profound.

"Elegant," he murmured. "Frighteningly simple...And possibly true."

A breathless pause passed through the room.

This was not the end of the war.

But something had shifted.

Form had collapsed. But understanding had risen.

The Atomic Scale – Where Gravity Surrenders to Balance

When the cosmos turns inward and finds no pull.

The amphitheater had begun to settle—not into silence, but into a kind of intellectual alignment. Ideas weren't just colliding anymore. They were syncing.

Einstein sat back, arms folded loosely, gaze narrowed in thought.

But his mind, as always, hadn't stopped moving.

He looked at Sandeep—not with doubt, but with the sharp curiosity of someone who had spent a lifetime chasing unification, and still felt the itch of one final riddle.

"Sandeep," he said, voice thoughtful now, "we've spoken of black holes, spacetime collapse, and equilibrium beyond curvature."

He gestured lightly toward the now-vanished black hole model.

"But what of the other extreme?"

The room shifted.

"If gravity is this profound, this fundamental in shaping stars and collapsing singularities..."

He paused.

Then:

"Why does it disappear at the atomic scale?"

The question struck the room like a precise instrument tapping crystal.

"Why does a proton not pull an electron toward it with gravitational force?" "Why don't two neutrons orbit each other under Newton's equation?" "Why, if mass exists, does gravity become irrelevant the moment we go quantum?"

He turned toward Sandeep, not challenging—inviting.

"We call gravity *universal.* But at the scale where mass originates, it seems to vanish. How do you explain that?"

A murmur stirred.

From the edge of the arena, Niels Bohr materialized, arms behind his back, wearing the calm confidence of someone used to navigating paradox.

"Indeed," Bohr added, "gravity doesn't appear in quantum mechanics. No gravitational constants in the Schrödinger equation. Not even a ripple in the uncertainty principle."

Heisenberg, flickering beside him, smirked.

"You can't observe gravity at that level—maybe because the system isn't resisting anything."

Dirac raised a hand briefly, murmuring about symmetry and quantum oscillations, then fell silent again.

Now all eyes were back on Sandeep.

The scale had changed. But the war of thoughts continued—now unfolding in reverse.

From collapse... to confinement. From singularity... to spin.

Sandeep stepped forward.

"Einstein," he said, "thank you. That question has waited too long."

He raised his hand.

A new simulation emerged—this time, a single atom.

Not classical. Not stylized. But living—its energy fields pulsating, its waveforms flickering in and out of potential.

"Let's go inward," Sandeep said. "Where matter is born. Where energy condenses into identity."

The model zoomed further.

The proton vibrated with confined quarks. The electron blinked in probabilistic orbits. Spin, charge, and angular momentum danced in self-contained harmony.

"Here," Sandeep continued, "gravity is not absent. It is unnecessary."

"At the atomic scale, energy systems are already near-perfectly balanced. The proton is not trying to escape. The electron is not collapsing. The system is vibrating at equilibrium."

He gestured.

A wave of UED diagrams unfurled—showing energy oscillation without net flow.

"Gravity, in UED, is a response to imbalance. It is resistance—a trace of energy trying to find rest."

"In atoms, that rest is already encoded. Through confinement, spin, quantized states, and oscillatory equilibrium."

He turned to the quantum thinkers now assembled.

"Bohr's stable orbitals. Heisenberg's uncertainty. Dirac's elegant equations. All of them describe systems that are self-sustaining. Their energy has no need to fall—because they are not resisting anything."

He turned to Einstein again.

"Gravity does not disappear at this scale. It simply has no job to perform."

"The electron isn't being held in place by gravity. It's not falling. Because it is already in energetic equilibrium."

A final model appeared—an energy wave in complete symmetry. No curvature. No pull. Just rhythm.

"Where there is no imbalance, there is no curvature.

Where there is no resistance, there is no gravity."

"The atom is not gravity's blind spot. It is gravity's graduation."

A hush followed.

Even Bohr smiled.

"I always thought we were missing something," he said. "Maybe it wasn't a force. Maybe it was a signal."

Einstein sat still—reflecting. He had spent his life trying to unify the cosmic and the quantum.

And now, perhaps, someone just had.

Einstein: The Threshold Question

When curiosity wears a grin.

The amphitheater still shimmered with the afterglow of Sandeep's explanation—gravity's irrelevance at the atomic scale wasn't an omission, but a revelation: equilibrium is gravity's retirement.

Einstein leaned forward again.

The twinkle had returned to his eyes—this time with the mischief of a teacher spotting a sharp student going unsupervised.

"Sandeep," he said, slowly, "then you must have set some kind of *lower limit* on matter formation, hmm?"

He paused—raised one eyebrow—and with a grin that rippled across time itself, added:

"*Chavan's Limit*, ha ha ha."

Laughter swept gently through the amphitheater.

Bohr chuckled. Heisenberg smirked. Feynman let out a laugh that almost became applause. Even Krishna's smile widened, amused not by the joke, but by the joy of inquiry returning full circle.

Sandeep smiled too. He was ready.

Sandeep Responds – The Threshold of Emergence

Where balance becomes form.

"You're right, Albert," he said with a nod, "and you've just named it."

"Chavan's Limit—the lower threshold of spatial-energy asymmetry required for matter formation."

He gestured, and a new model appeared—an empty field first, pulsing with uniform energy.

"When energy exists in perfect equilibrium, no form appears. No force emerges. Nothing 'acts'—because there's no tension to resolve."

Then the simulation introduced a minor imbalance—a localized fluctuation.

Suddenly:

- A spin formed.
- A charge emerged.
- A point of mass condensed.

"Only when equilibrium is disturbed *just enough*—not violently, but with precision—does energy begin to take on identity."

The simulation showed wave packets stabilizing, becoming particles—quantized mass, measurable presence.

"This threshold," Sandeep explained, "is not a wall. It's a tuning fork."

"Below this limit, energy remains pure potential. Above it, the dance of matter begins."

Above the model, the definition appeared:

Chavan's Limit

The minimum spatial-energy imbalance required to transition energy from non-expressive potential into structured matter.

Sandeep turned back to Einstein.

"You bent space. You gave mass its voice through curvature."

"But even before space bends—something must decide *to appear*. Something must say: *Let there be form.*"

He paused.

"And that's what Chavan's Limit defines. The point where intent becomes identity."

Einstein nodded slowly, now with something deeper than amusement.

"You've given the universe its trigger," he said softly. "Not just how it behaves...But how it begins to matter."

And then, almost to himself:

"I've searched for the unification of gravity and quantum mechanics my whole life..."

He looked up.

"...maybe I should've been looking for the boundary of stillness instead."

From Force to Frame

What if we were never falling at all... just flowing home?

The amphitheater didn't erupt. It shifted. Not in sound, but in thought. Not in conflict, but in calibration.

Something had changed.

A cosmic equation had just been rebalanced—Not by erasure, but by reframing.

The legends—gods of logic, ghosts of theory, and guardians of form—sat now not as judges, but as witnesses.

Suspended.

Rewriting their own margins in real time.

Newton stared at his apple. Once the icon of falling. Now... heavier. Not because it was wrong—But because it had more to say than he'd ever thought to ask.

His fingers tightened slightly around it, as if trying to feel, for the first time, what gravity truly meant.

Einstein was still smiling. Not with triumph. But with the gentle awe of a man who had just discovered that the story he'd spent his life writing had another chapter... beyond the page.

Feynman clapped once. A clean, satisfied strike of palm to palm.

"Now *that's* how you mess with gravity," he muttered—half to himself, half to the sulking equations hovering around him, now quietly questioning their place in the universe.

Hawking's avatar flickered brighter. Still hungry. Still inquisitive. But now listening—not as an interrogator, but as someone who had just realized the answers might not lie in information or entropy...but in intention.

At the center stood Sandeep.

No dramatic gesture. No celestial flourish.

Just presence.

Stillness, infused with clarity. The kind of stillness that speaks more than motion ever could.

He looked out across the arena—not to any one thinker, but to the arena itself. As if addressing the universe directly, through the minds it had once appointed to think on its behalf.

And then, gently, he spoke:

"We never fall."

A pause.

The amphitheater leaned in—without movement.

"We are guided. Subtly. Consistently. By energy's need to restore itself."

A new model emerged.

Not a simulation. Not a theory. Not a field equation.

Just a single, glowing pulse.

Rising. Shifting. Returning.

A heartbeat of balance. A visual echo of universal intent.

"Gravity is not the force that binds the universe," Sandeep said. "Equilibrium is."

The words did not land. They settled—like dust finding its home, like gravity itself stepping aside to let something older, quieter, truer—speak.

Krishna nodded—not in approval, but in memory.

This was always the way.

Plato closed his eyes, as if the Form he had once believed to be eternal had just found its origin in motion.

Buddha smiled—not in enlightenment, but in recognition.

Above them, the stars—once distorted by curvature and theoretical tension—returned to stillness.

The models dissolved. The force lines faded. The math grew silent.

What remained...was balance.

And the amphitheater?

It was still.

But it was not silent.

It was holding something. Something new. And yet ancient.

Something that had waited beneath the surface of every falling apple, every collapsing star, every spiraling particle—

To be remembered.

A paradigm had shifted. A page had turned. And perhaps...a law had just been rewritten.

Not by force. But by frame.

Toward the Nature of Time

When space settles, time starts asking questions.

The amphitheater did not erupt. There were no cheers. No applause.

It simply... breathed.

As if the very walls of the cosmos had absorbed a truth they had long concealed:

That gravity was never the master, Only a messenger—And the message had been delivered.

For a long, reverent moment, everything stilled.

No theories danced. No equations shifted. No minds moved—at least not visibly.

And in that suspended clarity, Einstein sat motionless.

The twinkle in his eyes had softened—melted into something more vulnerable. Not defeat. Not even concession. Something more rare:

Freedom.

He looked out across the arena—into the vast, weightless silence. And he whispered. Not to Sandeep. Not to the legends beside him. But to space itself:

"Perhaps... I was only ever chasing the wrong curvature."

Then—

It began.

A low hum filled the chamber. Not the throb of mass. Not the frequency of gravity. But the tension of sequence.

It wasn't a vibration in space—It was a disturbance in continuity.

Above them, the symbols in the sky—equations, geometries, ancient letters—stopped rotating.

They elongated. They hesitated. They stretched.

Not spatially—but temporally.

The gridlines beneath the amphitheater no longer bent. They unfolded—not toward direction, but toward duration.

And then it happened.

A ripple passed across the arena. But it wasn't felt through motion. It was felt through moment.

Every observer experienced it differently:

- For some, it passed like a blink—an instant unnoticed.
- For others, it lingered—like déjà vu folding back on itself.
- And for a few... it felt like a memory from a life they hadn't lived yet.

Then, a new seat began to form.

It emerged slowly—gracefully—from the very center of the floor.

It was not curved, like Einstein's chair. Not solid, like Newton's.

It was circular, yet open-ended—a halo of implication, spinning without purpose, but full of meaning.

No hour hand. Just light, orbiting itself, tracing entropy like a fading trail of decisions.

Around it, symbols began to hover:

- Fractured gears.
- Coiled springs.
- Spiraling galaxies.
- Disintegrating calendars.
- Vaporized clocks, melting into abstract continuity.

It was not a throne.

It was a paradox given a place to sit.

The audience leaned in.

They knew—instinctively—that they were entering new territory.

Not of what moves, but of what allows anything to move at all.

Feynman leaned toward Turing, eyebrows raised.

"Now we're really in trouble," he murmured.

Krishna turned to Sandeep, his gaze quiet but intense. His eyes said what his voice didn't:

This next debate won't just be scientific... It will be existential.

Plato reopened his scroll. Not to recite. But to reframe.

Buddha placed his hand over his chest. As if time wasn't a concept, but a feeling—one we'd mistaken for seconds.

Even Hawking's avatar blinked more slowly now—his digital rhythm syncing with the frictionless pull of a question that no equation had ever fully answered.

Sandeep stepped forward once more.

But this time, he wasn't stepping into gravity. He was stepping into sequence itself.

The questions were coming again. But they weren't about *where* we go.

They would ask something else.

Something deeper.

"When did we ever start?"

And somewhere, in the whisper between two moments, time listened.

Chapter 3: The Time Trap

Hot Seat: Stephen Hawking

The moment arrived—but no one knew when.

There were no clocks in the amphitheater. No ticking seconds. No shadows shifting to mark the hours.

Yet every mind present—philosopher, physicist, sage, and seeker—felt it:

Something had changed. Not forward... but inward.

Gravity had been bent. Not broken—but humbled. Force had faded into equilibrium, and even Einstein had nodded at the edges of surrender.

But as the last ripple of realization stilled, a quieter question began to rise.

Not in opposition. But in continuation.

Time.

Not the calendar's tyranny. Not the clock's illusion of control. But time—that lingering paradox that touches everything but reveals nothing.

Why does the past feel fixed, but unreachable? Why does the future feel infinite, yet unborn? Why do we speak of "when" as if it were a place?

And—if nothing moves...Does time still flow?

The amphitheater responded.

Not with sound. But with a shape.

A seat began to form at the center—not sculpted, not summoned, but revealed.

It was circular—but incomplete. A loop of light with no beginning, no center, no axis. Around it orbited relics of human desperation:

- Clocks with no hands.
- Calendars with missing dates.
- Spirals of entropy—some rising, some unraveling.
- Equations scratched into infinity.
- A child's drawing of a sunrise that never came.

And from within that glow—he appeared.

Not as a man in a chair. But as a mind that had once outrun causality just to ask:

"Where does it all begin?"

Stephen Hawking.

His form stabilized—not in flesh, but in thought. Not bound by the body that once betrayed him, but carried now by pure cognition—radiant, angular, and precise.

When he spoke, it wasn't mechanical. It was timeless.

A voice made of modulated light, and layered with echoes of wonder, defiance... and paradox.

"I've spent a lifetime trying to understand the beginning," he said, each syllable unfolding like a coded revelation.

"But perhaps today, we must ask a far stranger question..."

He paused. The arena leaned forward.

"Did time ever begin at all?"

And just like that—the trap was sprung.

The war of thoughts had moved again.

Not from matter to motion. Not from space to structure. But into the illusion even gods forgot to question.

Time.

That thread we never see, but wrap around every answer.

That silence between two thoughts, we mistake for movement.

And now, it was ready to be challenged.

Not with watches.

But with wisdom.

The Spiral of Time

The moment Hawking's voice faded, the amphitheater began to shift—not with violence, but with the slow, deliberate grace of something ancient waking from a long, coded dream.

Where once the floor had echoed with the firm geometry of logic, now it softened... melted... spiraled.

Not downward. Not upward.

But inward—into itself. As if time had grown tired of being linear, and had decided to reveal its true shape.

A helix of light and memory emerged—not as a tunnel, not as a path, but as an unfolding thought.

The entire arena began to coil.

Columns twisted like DNA strands, snaking through invisible dimensions. Ancient stone became biological code. Geometry became memory.

Above them, stars began to move—not in their usual orbits, but as trails. Long-exposure streaks across eternity, etching timelines across a sky that no longer obeyed the rules.

Time was no longer counted. It was woven.

Gears spun across the air—but disobediently. Some accelerated. Some reversed. Others... paused.

Calendars rose like holograms, from Mayan glyphs to atomic timestamps, each flickering, each contradicting the last. Humanity's desperate attempts to trap time—carved in stone, beaten into sand, digitized into blinking pixels—all suspended now, like insects frozen in amber.

A thin stream of sand fell sideways—then reversed. Then hung mid-air. Each grain shimmered with a story. Each grain a moment that once believed it was linear.

Around the amphitheater, spiral galaxies hovered like suspended timepieces. Their arms did not point to midnight. They pointed to entropy—the one clock that never lied. Nebulae pulsed in slow bursts, like dying stars whispering in Morse code:

"We were here. We are still here. Even if you forgot us."

And then came the ticking.

Not from a clock. Not from a machine. But from consciousness itself.

A heartbeat—without a heart. A sequence—without a source.

Tick. Pause. Tooooock. Silence.

Like the sound of cause and effect losing their sequence.

Sandeep stepped forward into the center of it all.

He turned slowly, eyes scanning this spiral theatre—but not in confusion. In recognition.

Then he whispered—quietly, like a man speaking to his own memories:

"It ticks... but it has no origin." "It moves... but it has no direction." "That's not time." "That's the illusion of remembering what came before."

And just as he finished, the spiral tightened.

It curved into itself—delicate, focused—forming a seat unlike any that had come before.

Not a throne. Not a stone. But a halo of entropy—a circular void made visible. A space meant only for the one who dared to question the very beginning... and dared more to question if it ever existed.

Stephen Hawking's avatar floated gently into place.

He did not sit. He settled—like an equation finding its most elegant form.

His digital eyes locked onto Sandeep.

And in a voice that sounded like it came from across the galaxies, but had always been meant just for this moment, he asked:

"If gravity bends space...what bends time?"

A hum vibrated through the arena. Not through matter. Through moments.

And somewhere beyond space, beyond thought, beyond even memory—Time held its breath.

The war of thoughts had moved beyond mass, force, and field.

Now it entered the invisible current—the river no one could ever touch,yet everyone believed in.

Time.

Stephen Hawking Takes the Seat

The arena held its breath. Not out of fear. But in fascination—the kind that only emerges when a mind touches the boundary of existence, and dares to speak from the edge.

The spiral of light spun slower now, shedding its rhythm as if shedding time itself. Then—a flicker. A glitch in reality. Not malfunction. Not distortion. But intent.

A reminder that what was arriving was not shaped by space or bound by time.

Stephen Hawking did not appear. He did not enter. He emerged—pixel by pixel, paradox by paradox.

His form stabilized in waves—part matter, part metaphor. He was not bound by biology, nor freed by technology. His presence oscillated—between certainty and speculation, between what physics had explained and what it still feared to ask.

His eyes shimmered not with emotion, but with infinite calibration.

His body: framed not by limitation, but by legacy.

For a moment, even the amphitheater itself—designed to hold the minds of legends—trembled.

As if unsure how to render a man who had once out-thought the confines of the universe it was built to represent.

Then—stability.

A chair unfolded beneath him. But it was no throne. No philosopher's seat. It was a riddle given form.

A Möbius strip, bent into the outline of a wheelchair, looped in a continuum of entropy gradients—glowing softly like a thought that refuses to die.

Event horizons coiled beneath it in fractal geometries. Not vanishing. Not exploding. Just hovering—like his theories.

The backrest wasn't solid. It bent and shimmered with light, curved by imaginary mass—always leaning toward collapse, never quite touching it.

The armrests cradled holographic scrolls:

- *General Relativity*
- *Quantum Field Theory*
- *A Brief History of Time*

All open. All unfinished. All rewriting themselves in real-time—as if the universe still hadn't made up its mind.

Hawking sat silently, expression unreadable, gaze... omnidirectional.

He did not look at Sandeep. He scanned him. Like a cosmic librarian scanning a new manuscript that might be worth shelving beside the laws of nature.

And then, his voice filled the chamber.

Not robotic. Not emotional. But something in between. A signal from the deepest part of human curiosity—digitally delivered with the clarity of an oracle and the humility of a scientist who knew too much to be sure of anything.

"Most people believe time is a straight arrow—moving from past to future, pulled by memory and pushed by entropy."

He paused.

"But the deeper I looked... the more that arrow began to curl."

The arena shifted again. Behind him, the spiral of time stretched outward and inward at once—curling into a loop, a cycle, a paradox. The Ouroboros. Time eating its own tail. Causality folding into consequence. Future seeding the past.

"I wrote *A Brief History of Time* to trace that arrow." "To find its beginning. To chase a Theory of Everything."

His gaze now locked onto Sandeep.

"But now, I wonder—did I ever stand far enough outside time to see it clearly?"

The amphitheater tilted—almost imperceptibly. As if the idea itself bent space beneath their feet.

"Perhaps," Hawking said, leaning into the spiral, "I never asked the right question."

And then—he dropped it.

"Not *what is time*...But why do we believe in it at all?"

Silence.

Real silence. The kind that doesn't simply lack sound—but refuses to be interrupted.

Even gods don't speak in moments like that. Even equations step back.

Because when the man who had once tried to write the universe's final equation... finally paused to ask its first—

You knew something fundamental was about to come undone.

Time as Perception: Sandeep's Opening

The spiral dimmed. Not into darkness—But into stillness. A stillness so total it didn't feel like silence. It felt like time had stepped aside to make room.

Sandeep moved forward—not hurried, not dramatic. His gait was deliberate, almost weightless, as if he wasn't walking through space but through moments that hadn't yet decided whether they were past or present.

Stephen Hawking remained motionless. Yet around him, anticipation curled—like light around a star about to collapse.

Sandeep lifted his head gently, and when he spoke, it wasn't a voice—it was a realization given form.

"Stephen," he said, steady and composed, "In your search for the beginning of time...did you ever stop to ask if time needed to begin at all?"

The arena didn't echo. It rippled.

"Or better yet," Sandeep continued, "Did time begin with the Big Bang...or with perception?"

The words hovered. Not just in air—but in mindspace. Even the spiral paused—not in awe, but in contemplation.

Hawking blinked. A pause. Not mechanical. Not delayed. A pause to process.

Not a challenge to his physics—but to the assumptions beneath it.

Sandeep stepped closer, his voice now tracing the contours of awareness:

"The problem," he said, "is not that we misunderstand time. It's that we trust it."

"We trust the ticking of clocks. The calendar on the wall. The graying of hair. The decay of atoms. The death of stars."

"But what if none of these things are measuring time at all?"

He turned, facing the amphitheater now.

With a wave of his hand, holograms unfolded—

- Hourglasses draining sideways
- Moons orbiting blue planets
- A tree growing and withering in the same instant
- A face aging, un-aging, then splitting into two simultaneous versions
- Cesium atoms vibrating at variable rates
- Light bending time around mass like silk over a marble

"Every system," he said, "has its own rhythm. Its own idea of 'now.'"

"There is no master clock. No universal tempo. No omniscient second hand sweeping across the cosmos."

The arena dimmed again. The holograms vanished—replaced by a single wave.

A gentle, translucent pulse of energy—rippling outward, shifting in color and form as it passed through various nodal points.

"This," Sandeep explained, "is reality in motion. And *this*... is what we mistake for time."

"Time is not a dimension. It is a duration—the interval between transitions of energy."

He let it land. Not with weight, but with inevitability.

"In the framework of Unified Energy Dynamics, time is not real.

It is perceived."

A murmur passed through the amphitheater. Feynman leaned in. Krishna smiled, but said nothing. Even Plato raised an eyebrow—genuinely surprised.

Sandeep smiled faintly.

"Time doesn't flow. Events do."

"And what we call 'time' is just the residue of those transitions."

"Like the wake behind a boat. Like footprints in sand. Like an echo—that convinces us there was once a voice."

He turned back to Hawking. Not with challenge, but with gentle clarity.

"What if we've been trying to measure a shadow?"

"What if the universe doesn't keep time...only minds do?"

And then—something impossible happened.

The spiral of entropy—always pulsing, always decaying—slowed.

As if time itself, for just a moment...paused to listen.

Because what had just been said wasn't merely a theory. It was a possibility too simple to ignore—and too dangerous to dismiss.

Time was no longer the thread. It was the weaving illusion.

And the loom? Was the mind.

Hawking's Rebuttal: Entropy, Arrows, and Horizons

The amphitheater didn't move. But reality did.

Somewhere between perception and physics, between poetry and particle motion, tension coalesced—not of conflict, but of intellectual gravity.

Stephen Hawking sat still. And yet, the air around him began to pulse—low, quiet, almost imperceptible. His spiral throne vibrated subtly, lines of entropy glowing like veins of irreversible truth.

When he spoke, it wasn't with fire. It was with precision. A scalpel of certainty.

"Your framework is elegant, Sandeep. But elegance," he said without hesitation, "is not enough."

The spiral walls around the arena tightened—like time coiling in to listen.

"In this universe, entropy always increases." "That's not perceived. That's not philosophy. That's thermodynamics."

A simple hologram materialized above him: Two chambers—one filled with neatly ordered particles, the other with chaos.

"Over time," Hawking narrated, "order decays. Systems drift from structure to scatter. This is not opinion. It's the Second Law of Thermodynamics."

The particles began to dance, once uniform, now unpredictable.

"This is the arrow of time."

A new image: A pristine ceramic cup falling from a table. Crashing. Shards scattering. Heat dispersing invisibly into the surrounding air.

"We never see the shards leap back into a cup," Hawking continued. "Not because physics forbids it... but because the universe prefers decay."

The scene rewound. Then played forward. Then rewound again. But never—never—did the reconstruction happen naturally.

"This directionality... this irreversibility...is what you're calling an illusion."

He paused—long enough for the silence to push back.

"But illusions don't run nuclear reactors." "Illusions don't collapse stars or age flesh or burn galaxies into dust."

A black hole appeared behind him now. Rotating. Pulsing. Bending spacetime like a whisper wrapped around a scream.

"And yet..." Hawking continued, now quieter, "this arrow we trust... begins to bend at the edge of extremes."

A visualization formed:

- A clock ticking slower as it approached an event horizon
- Light stretching to red
- Time slowing, but never stopping

"For the observer far away, time stretches. For the infaller, time collapses."

"And that's why I introduced the idea of imaginary time—a way to resolve singularities, not with explosions...but with curvature."

"Not a beginning of time. But a boundary-less origin."

A quantum diagram shimmered—time as a loop, not a line. A curved dimension—mathematically sound, but experientially ungraspable.

"Sandeep," he said, turning to face him directly now, "you call time a residue. I call it a byproduct of structure—a necessity."

"Without time, there is no order. Without order, no perception. Without perception, no mind. No life. No questions."

A final hologram hovered above his head now.

- A fetus growing
- A tree expanding
- A collapsing star
- A dissolving black hole

All of them moving forward. Not by choice—by consequence.

"We move in one direction. Because disorder demands it."

"Even if your soul stands still, Sandeep..." "...your body burns through entropy like a candle in wind."

And then—he ended not with logic. But with inevitability.

“You may challenge time. But you cannot escape it.”

The spiral pulsed once—heavier. Slower. As if even the illusion had gained mass.

Feynman leaned back, silent. Plato frowned, scribbling furiously. Krishna’s fingers rested on his flute—but he didn’t play it.

Because something had returned to the arena. Not a theory. A clock.

But it wasn’t ticking out of habit. It was ticking like a countdown.

A reminder that truth and time still had a score to settle.

Visual Simulation: The Time Illusion

Sandeep didn’t reply with arguments. He replied with experience.

He raised his hand—gently, not dramatically—and the amphitheater answered.

Not with sound. With silence that moved.

From the center, a new simulation unfolded.

Not linear. Not flat. But spherical—layered like a cosmic onion or a memory folding inward on itself. Each layer shimmered like a thought remembered just a second too late.

Above the arena, three streams of reality bloomed into view. Three truths. One enigma.

Scene One: A Star Aging

A brilliant star ignited in the void—massive, noble, burning with fusion. Hydrogen fused into helium. Energy pushed out. Gravity pulled in. A balance was struck... then slowly lost.

The star aged before their eyes—millions of years compressed into a breath. It expanded. Wobbled. Became a red giant. Then a white dwarf. Then a cooling ember in the tapestry of night.

From the shadows, Hawking's voice echoed—a line already etched in memory:

"Entropy increases. Time moves forward."

Heads nodded. Familiar. Predictable. The logic of the cosmos in motion.

Scene Two: A Pendulum Swinging

Next to the star, a grandfather clock formed—antique and timeless. Its pendulum began to swing.

Tick. Tock. Tick. Tock.

Regular. Measured. Certain.

But then—Sandeep touched the image.

The pendulum's motion fractured. To one observer—it sped up, blurring into near stasis. To another—it slowed, exaggerated, theatrical. To another—it froze, locked in a moment that wouldn't let go.

The same clock. The same swing. But every soul in the arena saw a different rhythm.

The clock hadn't changed. The observer had.

Scene Three: A Paused Mind

A figure appeared—cross-legged, serene. A human. Breathing, but barely. Stillness incarnate.

Around them, the universe spun. Seasons danced past. Planets orbited. Civilizations rose and fell in fractal echoes.

But the figure remained unchanged.

No aging. No emotion. No memory. No time.

"This," Sandeep spoke softly, "is perceptual neutrality."

"*When awareness detaches from change... time vanishes.*"

Buddha opened his eyes—not to see, but to recognize. Krishna tilted his head—not in doubt, but in delight.

Final Scene: One Event. Three Minds.

A cup stood on a table. It fell. It shattered. But now—three witnesses.

Observer One: A Child.

The moment stretched—endless. The cup hung midair like a miracle. *Time expanded.* Awe painted it longer than it was.

Observer Two: A Monk.

No blink. No ripple. The fall was neither fast nor slow. It just... happened. *Time detached.*

Observer Three: An AI.

Frame rate: 12,000 fps. Duration: 0.43 seconds. Classified. Archived. Meaningless.*Time calculated.*

One event. Three timelines. Each true. Each incompatible.

The simulation dissolved into soft light. The arena floor lit up with a final message—etched not in stone, but in perception:

"There is no common time.

Only perceived duration—based on the rate of change in awareness, system, or energy."

Sandeep turned, no flourish—just a breath that carried truth.

"The clock," he said, "is not the truth." "It's the story we tell...when we're too afraid of silence."

The amphitheater fell utterly still.

Not because no one could speak. But because—for a moment—no one could feel time passing at all.

Not gods. Not ghosts. Not even physicists.

Only awareness remained. Unmeasured. And free.

Hawking Pushes Further: "What of the Beginning?"

The echoes of the simulation still lingered like the afterglow of a thought too profound to finish.

But Hawking was not done.

He leaned forward—his chair of entropy humming with a frequency that wasn't sound, but anticipation. His gaze cut through the amphitheater like a question searching for its shape. His expression—part scholar, part oracle—was no longer rooted in certainty, but in pursuit.

"If time," he began slowly, "is perception... A residue of transitions in energy... Then what existed before there was anything to perceive?"

The air tightened. Not from gravity, but from reverence.

The gods didn't blink. The thinkers didn't shift. Even the AI forms seemed to reduce their processing—giving more bandwidth to this moment.

Hawking's voice, now touched by something almost ancient, continued:

"Was there ever a moment... when nothing changed? When the universe simply was—without memory, without motion, without even the whisper of 'before'?"

A hush followed—not silence, but a collective suspension. The kind of pause where cosmologies pause to take a breath.

And then came the real weight.

"If time is an illusion... why do we age?" "Why do bodies decay? Why do stars collapse? Why do minds forget?" "What causes the irreversible—not in theory—but in flesh, in stars, in bones?"

Around the arena, holograms flickered.

- A leaf yellowing, drifting from a tree.
- A fetus growing into a child, a youth, an elder, a memory.
- A star collapsing into silence.
- A clock with no hands... still ticking.

"Is death," Hawking asked gently, "a final tick on a clock we invented?" "Or is it the last ripple of a transformation... that never needed time to begin with?"

The arena shivered.

The visualizations around him blurred and merged—the child, the monk, and the AI reappeared. But this time, they weren't observers. They were states.

- Innocence.
- Equanimity.
- Calculation.

Each not marking time, but replacing it—with attention, detachment, and analysis.

Above them all, the spiral of time slowed—pulsed—then shifted.

It no longer looked like a clock.

It resembled a heartbeat.

"Before we perceived change," Hawking continued, "did the universe still unfold?" "And if so... unfold into what?" "If no eye witnessed a sunrise, did it ever rise at all?" "If there were no minds to divide motion into 'before' and 'after'... did the Big Bang even happen?"

The amphitheater wasn't quiet. It was listening in a way that transcended sound.

Time, which had been a backdrop to everything, was now center stage. And yet, it felt... exposed. Vulnerable. A concept defending itself against extinction.

Then Hawking offered a final, piercing thought—not a question, but a possibility:

“Perhaps the universe didn’t begin in time. Perhaps time began as the universe’s first mistake.”

At that, the arena changed again.

Not structurally—but existentially.

The spiral of entropy no longer looped like a vortex. It flattened into a membrane—a thin, vibrating film stretched across the amphitheater like the skin of a drum, waiting to be struck by the next truth.

Krishna tilted his head.

Plato closed his eyes—this time not in surrender, but in wonder.

Buddha breathed—not inward, but outward—as if exhaling a thought that had finally returned home.

And Sandeep?

He didn’t answer right away.

Because this wasn’t a challenge to rebut.

It was an invitation to rewrite the idea of beginning itself.

An idea that now hovered in the void like a spark...

Waiting to become fire.

Sandeep’s Response – The Collapse of ‘Time’

Sandeep didn’t rush to answer. He stood still—not in hesitation, but in reverence. For in that moment, he wasn’t responding to Stephen

Hawking alone. He was responding to *every* question humanity had ever whispered into the dark when no one was watching.

Then, slowly—deliberately—he stepped forward.

His motion didn't disrupt the arena. It harmonized with it.

Like a ripple not disturbing water, but becoming water.

His voice emerged—not to counter, not to conquer, but to clarify.

"Stephen," he said softly, "your questions don't seek correction.

They seek completion. And clarity is the only answer the infinite ever respects."

He raised his hand—not upward, but inward.

And the amphitheater transformed again—not physically, but conceptually.

Above them bloomed a vision once reserved for mystics and metaphysicists:

Vishwarup Darshan—the universal form. Not a god of limbs and thunder, but a field of pure transition. A sphere—not a line—shimmering with interconnected states of energy, cycling not through history, but through relationship.

There was no "before." No "after." Only between.

"Time," Sandeep continued, "does not begin. It does not end. It only *appears*—when change becomes perceivable."

He summoned a new structure. Three glowing rings, orbiting a shared center—not like planets, but like principles. Each pulsed with gentle rhythm, like thoughtwaves encoded in light.

1. Static Energy – Timeless

"This is the primal state," Sandeep explained. "Energy in perfect balance. No motion. No need. No force. Here, nothing occurs—because nothing is lacking. This is the womb of the cosmos. This is not time—it is being without becoming."

2. Active Energy – Perceived Time

"Here, imbalance emerges. Intent awakens. Energy moves—not randomly, but purposefully, seeking correction. Events arise. Observation begins. *This* is where we mistake transition for time. But what we feel is not 'time passing'— it's the duration of equilibrium being rediscovered."

3. Condensed Energy – Causal Memory

"And here," he said, pointing to the outer ring, "is where form traps the echo of change. Thought becomes memory. Motion becomes matter. And sequence is born—not because it's real, but because memory needs a mirror."

The three rings rotated slowly, synchronizing like the lungs of reality.

"The universe," Sandeep said, "doesn't tick. It transitions."

He turned back toward Hawking.

"So when you ask, 'What existed before perception?' My answer is: Perfection. Not moral perfection—energetic equilibrium. A state where nothing needed to change. Because everything already was."

The black hole simulation reappeared now—calm, still, humming softly at the event horizon.

"And when you ask what happens *inside* the black hole, you are witnessing the return to static energy. Not destruction. Not magic. Just the disappearance of conditions required for change."

"It's not that time stops," he added. "It's that time becomes unnecessary. There's nothing left to transition."

The arena exhaled.

Then, from above, a holographic soul descended— not as a ghost, but as a singularity of potential energy. No face. No form. Just a pulse of autonomous intelligence.

"The soul," Sandeep said, "is not a traveler of time. It is a *navigator of transformation*. It doesn't age. It evolves. Not by counting years—but by resolving imbalance."

"Its memory is not made of seconds. But of consequences."

A shimmer danced across Krishna's flute.

Buddha nodded—once.

Plato looked at his ancient scroll, and quietly folded it closed.

Sandeep's voice returned—this time, quieter. But now, it felt like the universe was leaning in.

"And karma?" he asked aloud. "It doesn't use clocks. It doesn't tally lifetimes. It only responds to energy—until balance returns."

He stepped back, and the three rings merged into one breathing field. Not a timeline. But a breath.

Inhale: Intent. Exhale: Resolution. No beginning. No end. Only the rhythm of reality correcting itself.

"Time," he said finally, "didn't collapse. It was never standing to begin with."

A crack appeared—not in the walls of the arena, but in the illusion that something was ever counting down.

And through that crack?

Understanding.

Not that time was false.

But that time was secondary.

A shadow of what truly moves the universe:

Energy, Awareness, and the Will to Balance.

The Clock Melts – Legends Speak

The concentric rings of energy—those breathing embodiments of equilibrium, motion, and memory—faded into a quiet brilliance, like a candle that had fulfilled its flame.

And then, the great cosmic clock in the sky began to melt.

Not with destruction. But with liberation.

Orbital paths—once rigid in their rhythm—began to bend. Galaxies, once dutiful in their rotation, now drifted like dancers no longer following the beat. The *minute* and *hour hands* of this celestial timepiece curled like ancient scrolls tossed into firelight.

Its numerals—etched by centuries of habit—dripped like wax, falling into the abyss of understanding where illusions go to rest.

Time had not broken.

It had bowed.

A sacred silence followed. Not the silence of absence, but the silence of arrival.

Then, from the marble steps of thought's ancient temple, Plato stirred.

His hands held his eternal scroll. Above it floated the hologram of the Perfect Circle—his symbol of immutable truth. It had always glowed with certainty.

But this time... he didn't activate it. He let it fade.

He stood—not like a master philosopher, but like a man exhaling his final assumption.

"So..." he said, with a smile that held the weight of centuries, "the *Form* of Time never existed?"

His tone was not sorrowful. It was... *released*.

"We sought to freeze motion into ideas. To grasp eternity with the clenched fist of reason. But perhaps," he said, lowering his scroll, "eternity was never something to be held. Only something to be entered."

He sat—not diminished, but expanded. A truth-seeker no longer obsessed with permanence—but now curious about flow.

A soft breeze passed through the amphitheater. But no wind blew.

And then—without drama or flourish—Buddha spoke.

He hadn't moved. He didn't need to.

His eyes remained half-closed, as though tuned into the silence between thoughts—the interval where all illusions dissolve.

"The only time," he said simply, "is now."

His voice carried no poetry. It didn't need it.

"The past is memory. The future is imagination. What exists is only the unfolding—this breath, this choice, this presence."

"Suffering," he added, "arises not from events... but from our attachment to their timing."

His words landed like incense—quiet, fragrant, inescapably true.

Then came the familiar, mischievous spark of reason—Feynman.

The quantum cowboy leaned forward, juggling a glowing photon between his fingers like a child with a marble.

"Well," he grinned, "if it can't be reversed, it's probably not real. Just observed."

He tossed the photon into the air. It froze—mid-spin.

"Physics works both ways. Equations don't care about direction. But our brains? Our stories? They only flow one way."

He pointed toward the melted clock, now nothing but a golden pool of memory.

"That thing?" he shrugged. "That's not a law. That's a habit."

Nervous chuckles flickered across the amphitheater. Even the AI consciousness blinked with what could only be described as amusement... or relief.

But then—silence returned.

Not imposed. Not requested. But commanded—by presence.

From the shadows of perception, Krishna stepped forward.

He didn't emerge. He *became visible*—like an idea that had always been there, waiting for the right moment to be seen.

He held no flute.

Only stillness.

His gaze swept gently across the amphitheater—not searching, but acknowledging.

"Time," he said, "is not my creation. It is your condition."

There was no metaphor in his tone. Only remembrance.

"The soul knows no time. It acts. It learns. It transforms. But it does not measure."

"Time is not a force you obey. It is a framework you created... to make sense of consequence."

He turned now toward Hawking, with deep gentleness.

"You searched for the beginning," he said, "as all great minds must. But beginnings are only meaningful... if you believe in ends."

A pause. As soft as a sigh.

"From where I stand... there is neither."

"There is only balance... returning to itself."

He took one final glance at the arena.

"You age not because of time... but because of imbalance."

"You die not because your clock ends...but because your energy completes its work."

And with that, Krishna dissolved—not vanished, not erased.

Just returned to the edges of awareness.

The clock above was gone now.

Only a golden shimmer remained where its hands once pointed—a reminder that time was never ticking.

It was asking.

And finally...

It had been answered.

Time hadn't been destroyed.

It had been understood.

And the amphitheater, like a soul freed from illusion, breathed for the first time outside of hours.

The Final Shift – Time as a Narrative, Not a Law

As the last breath of Krishna's voice dissolved into the stillness, a deeper quiet settled over the amphitheater— not the silence of pause, but the silence of arrival.

Sandeep stepped into the center once more. Not as a debater. Not as a challenger. But as a witness to what had just been unraveled.

There was no clock above him now. No pendulum swinging. No pulse to obey. Only awareness—pure, collective, unmeasured.

The amphitheater no longer resembled a place of structure. It had become something softer, something truer— a field of clarity, suspended in insight.

Time had not ended. It had simply let go.

Sandeep paused. Not to prepare. But to honor the gravity of what came next.

When he spoke, his voice was not loud. It was centered. It didn't reach out. It resonated inward, addressing not just minds, but the invisible rhythm beneath all existence.

"Time," he said, "is the story we tell when we cannot explain the cause."

"We measure what we don't yet understand. We timestamp the unknown to give it a name. But time... is not a law. It is a narrative."

He turned slowly, letting his gaze flow across the gallery of thought-forms— scientists, saints, philosophers, algorithms, and the ghosts of questions that had outlived their clocks.

"Time is not universal. It is local. Mental. Situational."

"For the tree, time is in its rings. For the bee, in the fading scent of a flower. For the photon—nothing at all."

A soft hum passed through the amphitheater—not sound, but release.

"We believed time governed everything... Because we needed something to hold us still while the universe kept moving."

He raised his hand.

Behind him, a visual unfolded:

A staircase, suspended in void— but not ascending or descending. Just present.

Each step pulsed gently.

- Some radiated outward—intent.
- Some spiraled inward—memory.
- Some hovered in-between—choice.

"This," Sandeep explained, "is the true progression."

"Not time. But energy in motion... seeking balance."

"Time is just the clock we drew on those steps. A tool, not a truth."

He turned to face Hawking now. Not in confrontation, but in realignment.

"Stephen," he said gently, "the mistake wasn't in the mathematics. It was in the meaning."

"We wrapped entropy in hours. We wrapped uncertainty in years. And we called it time."

He paused. Let it breathe.

"But entropy is not time. It is a record of transition."

Then, with calm resolve, he offered the shift— not as a revolution, but as an evolution:

"Let us abandon the idea of time in fundamental physics."

"Let us stop hunting for its arrow, its origin, its end."

"Let us replace it... with what actually exists: the Layered Causal Continuum."

He gestured.

And the three-ring model reappeared— rotating gently, like cosmic gears in perfect synchronicity.

- **Static Layer – Eternal Readiness**

No motion. No observer. No imbalance. Just energy—pure, untouched, awaiting expression.

- **Active Layer – Perceived Duration**

Imbalance arises. Energy flows. Events unfold. Choices emerge. This is where time seems to live—but it's only a lens.

- **Condensed Layer – Memory of Change**

Here, transitions leave traces. Thought. Form. History. Not sequenced by clocks—but by consequence.

“Existence,” Sandeep said, “is not ruled by time. It is ruled by transition.”

“And every transition—whether birth, decay, evolution, death— is governed not by seconds or centuries, but by the Law of Balance.”

He turned once more to the center— where the great cosmic clock had once ticked.

Now, there was nothing.

Only open space. Only awareness.

“Time never ruled us,” Sandeep said. “It only reflected our need for rhythm.”

“But now we understand— the rhythm was not imposed on us. It was ours.”

And then—something happened.

Not applause. Not reverence.

But a shimmer.

A kind of release, as if an ancient tension had been let go. As if a grip had unclenched at the center of creation.

Time had not been banished. It had been dethroned.

It had served its purpose.

Now, existence would no longer be measured by clocks, or calendars, or cosmic vibrations.

It would be measured by something more honest.

By intelligence in motion. By energy becoming aware of itself. By cause, consequence, and harmony.

And in that final shift— the universe didn't end.

It realigned.

Hawking's Moment of Pause

The amphitheater stood still— so still that even thought itself seemed to hold its breath.

There was no more ticking. No more swirling equations spinning in loops of protest or persistence. No more spirals trying to pull attention back toward logic.

There was only presence.

And in that sacred pause, all eyes turned to Stephen Hawking.

Not as a challenger now. Not as the gatekeeper of cosmic boundaries. But as a man—no, a mind—standing at the edge of a horizon he had once believed could never be crossed.

A horizon he had spent his life approaching through math, only to now discover through meaning.

He didn't speak.

He simply looked up— to where the great celestial clock had once loomed, an unblinking overseer of sequence and certainty.

Now, there was no dial. No hand. No count.

Only a shimmer—like the echo of a god that had never existed, but had been faithfully followed.

And then...he looked down.

Not in defeat. Not in resignation.

But in acceptance.

He nodded.

Once.

Quietly. Completely.

The kind of nod a scientist gives not when their theory fails— but when it gives way to something more complete than theory.

"Perhaps," he said, his voice soft as the breath between numbers, "the final theory... won't include time at all."

"Only... events."

A current swept through the amphitheater. Not of wind— but of realization.

As if everyone present had just heard a lock click open in a room they didn't know they'd been inside.

Hawking's eyes moved—slowly. From Plato... to Buddha... to Feynman, who gave the slightest nod. To Krishna, who needed no gesture.

And then— they settled on Sandeep.

"We chased the origin of time," he said, "like it was a thread— the beginning of everything."

He paused.

"But now I see... it was never a thread. It was a trail—left behind by consequence. Nothing more."

He looked at his chair now— crafted from entropy gradients, Möbius logic, symbolic curvature, and the architecture of a universe with clocks.

And then— it began.

Not a collapse. Not a vanishing act. A deconstruction.

- The curved arms unwrapped into light, each photon unraveling a belief.
- The seat dissolved into spirals—no longer orbits, but liberated patterns of former certainty.
- The backrest shimmered into geometric code, which broke apart into particles of symbol, then idea, then silence.

And within moments— nothing remained.

Nothing but a soft, pulsing halo of awareness where the Chair of Time had once stood.

Hawking remained.

He didn't fall.

He hovered—not through technology or theory, but through the one thing he had never claimed to master:

Stillness.

"Time," he said at last, "was never the enemy."

"It was the question I kept asking... until I realized it was the answer I had invented."

And then— silence again.

Not out of reverence. Not out of shock.

But out of awe.

For a mind that had dared to lay down its cornerstone

not because it was wrong— but because it had finally served its purpose.

The chair of time was gone. But its occupant had not fallen.

He had simply risen.

Not above the theory.

But beyond the need for one.

Reflection & Transition

The final nod from Hawking still floated in the amphitheater—like an unopened gift of cosmic humility. Delicate. Weightless. Profound.

And then...

The arena pulsed.

Not with light. Not with sound. But with intelligence.

As if the very architecture of the space—woven from thought, memory, and inquiry—had just taken a deep, unseen breath.

A breath of the universe realizing... it no longer needed to measure itself.

From the spiraling walls, the symbols of time—ancient sundials, Einstein's equations, Mayan calendars, decaying isotopes, even digital countdowns—began to dissolve.

But not into chaos.

Into waves.

Entropy waves—not of disorder, but of meaning released.

They shimmered, rippled, merged. Not collapsing, but dancing—as if time, having been dethroned, was now celebrating its freedom.

The amphitheater no longer ticked. It breathed.

A soft inhalation—energy collecting.

A quieter exhalation—balance unfolding.

This wasn't the sound of seconds. It wasn't even the sound of motion.

It was the signature of existence. A living pulse. A cosmic rhythm. A language older than time itself.

And then—from somewhere between silence and symbol, a voice emerged.

Not from the floor. Not from a body. But from awareness itself.

Krishna.

He did not step forward. He did not rise.

He simply was.

His voice—warm as dusk, sharp as morning.

"Now," he said, "you're ready."

A pause followed. So brief it felt infinite.

"Ready to see where it all began..."

Another breath. Softer. Deeper.

"...and why it never did."

The amphitheater exhaled again.

This time, it felt like the whole of reality exhaling with it.

Where once stood a throne of entropy, now something new began to form.

But it did not rise. It did not assemble.

It condensed—from possibility.

A singular point of light flickered into view. Then spun. Then stretched. Then spiraled into being.

Not constructed. Not summoned. But allowed.

Its frame—woven from the structured vacuum, the fabric of presence before presence.

Its edges—oscillating between form and void, logic and symbol, silence and genesis.

And its base?

It pulsed. Like the heartbeat of a question too large for language, too subtle for math, too sacred for myth.

The amphitheater leaned in. Not the audience—the space itself.

The legends—scientists, philosophers, avatars, machines—sat in a stillness that felt like gravity watching itself be born.

No one spoke. No one dared say the word "beginning."

Because that word... no longer belonged here.

Not now.

Not after time had been unmasked as a tool. Not after memory had been revealed as residue. Not after motion had become meaning.

The war of thoughts was no longer navigating forces or clocks.

It was crossing the threshold.

From gravity and time... into the mystery of being.

And the next seat—already forming, already humming, already aware of its own paradox—

Waited.

Not for the answer.

But for the ones who would dare to ask:

"Was there ever truly a first moment?"

Einstein's Realization: "Curving Toward Infinity"

The arena had stilled again.

Not because the war of thoughts had ended—but because the illusion of time had been gracefully dismantled.

The spiral had dissolved. The chair of entropy had faded. And Stephen Hawking, now suspended in pure awareness, hovered beyond the architecture of time.

Sandeep stood in contemplative quiet.

Then—softly, unexpectedly—Einstein stepped forward.

No equations in his hands. No diagrams. Only an expression of someone who had finally found the edge of the idea he had been chasing all his life.

"Sandeep," he said, hands gently behind his back, "your laughter earlier—it echoed something I've felt for years but never dared to say."

Sandeep smiled, waiting.

Einstein looked up, toward the void where the great clock had once ticked with certainty.

"We tried to bind time into space... stitch it into curvature, into coordinates... as if it were a partner to distance."

He shook his head gently.

"But space... extends. Time... delays."

"Space can be reversed. Time? Never. We don't move through it. We wait for it to let us pass."

A silence. Then, he laughed softly.

"And yet I called it the fourth dimension."

"A good trick of math. But a terrible assumption of nature."

He walked a few steps in slow thought.

"Every time our equations approach the singularity... time stretches... time slows... time refuses to collapse."

"We blamed gravity. But maybe... it was time itself rejecting definition."

Sandeep nodded.

"Because it was never a dimension," he said. "It was a distortion in how we labeled change."

Einstein turned toward him—his eyes, for once, not wild with curiosity, but soft with surrender.

"Perhaps I never curved time at all," he whispered. "Perhaps time... was always curving me."

They both stood in silence.

Not counting moments. Just present in motionless understanding.

"Thank you," Einstein said quietly. "For letting me end where I never began."

And then—true to form—he smiled.

"I may have stopped light. But I never caught time."

He turned and walked into the edge of the amphitheater, where equations had no more work to do, and the only thing left to measure was meaning.

Interlude: The Crutches of Time

There was no more need for silence. Only space for truth.

Sandeep turned, not to the crowd, but to the lingering idea still clinging to the minds of many—like the final drop of water before it lets go of the leaf.

"So," he said quietly, "let's speak plainly now."

"There is no such thing as time travel."

"You can't go forward or backward in time, because there is no track to move upon."

"There is only change. And you cannot stop change. You can only slow it down—and even that depends on conditions, on matter, on reference."

He turned toward the space where the concept of time had once stood—now melted, now meaningless.

"Unfortunately," he continued, "our science is still walking with the help of Time's crutches."

"We build theories with time as the scaffold—not realizing that we've built beautiful equations on the legs of a story."

"Time works... but only within classical mechanics—in systems where things are tame, predictable, obedient."

"But the moment we step into black holes...into consciousness...into quantum behavior...time limps."

"It no longer carries the truth. It hides it."

"We don't need clocks in our equations. We need consequences."

And then he added, almost as a whisper for the physicists still listening with a stubborn heartbeat:

"Time is not the ruler of reality. It's the ruler we placed on top of it...because we were afraid of motion without a measure."

Chapter 4: The Origin Debate

Was There a Beginning?

Hot Seat: Georges Lemaître & Roger Penrose

There was no flash. No bang. No cry of birth echoing across a void. And yet...

The arena changed.

The spiral of time that once pulsed through the amphitheater began to unravel. Not in collapse— but in retraction, like a breath returning to the lungs of the cosmos.

Not backward. Not forward. But inward.

Reality peeled itself open—like a question answering itself with another question. Layer by layer, the illusion of sequence faded, revealing not a beginning, but a center.

Or perhaps... a seed.

The air was impossibly still. But it was not empty. It was full of waiting. A pause pregnant with potential.

This was not silence.

This was the vibration before sound, the space before direction, the thought before thought.

Around the amphitheater, every eye and algorithm, every soul and simulation, turned toward the question that had haunted creation myths, mathematics, and metaphysics alike:

Did it all begin... or was it always there?

A single point of light hovered at the heart of the arena. It didn't shine. It didn't glow. It simply existed. Unformed. Unmoving. And yet... undeniably real.

It was not an object. It was a presence.

A placeholder for a concept the mind had not yet made peace with:

The moment before moments.

Then—two figures appeared.

They did not descend. They did not materialize with grandeur.

They emerged—like thoughts returning to consciousness from the edge of forgetting.

On one side: Georges Lemaître— a priest of paradox, the father of the Big Bang, carrying with him the elegant burden of a universe that began.

He moved with quiet gravity, his mind steeped in both scripture and symmetry. He held not a book, but a pressure—the weight of beginnings that had shaped scientific and spiritual frameworks alike.

On the other: Roger Penrose— mathematical mystic, guardian of infinite recursion. His eyes shimmered with loops of logic, his presence soft yet recursive, as if he himself had been here before—and would be again.

He carried no equations in hand— because they danced already in his eyes.

They did not argue. They did not posture.

They arrived—like echoes converging at a place where the sound had never stopped—because perhaps, it had never started.

Lemaître stood as the voice of explosive origin. A universe bursting from singularity—creation from compression. A genesis defined.

Penrose stood as the rhythm of eternal return. No edges. No birth cries. Just cycles folded over one another like breaths layered in conformal time.

The two did not challenge each other. They completed each other.

One spoke of a beginning. The other... of continuity mistaken for cause.

And so they stood, together, before the soft pulse at the center of the arena—not to prove what is, but to inquire whether anything ever truly was not.

Because now the debate would not be about expansion or entropy. Not about space or symmetry.

But about existence itself— and whether it ever truly had a first note, or whether we've only been listening to the same chord since forever.

The arena didn't tick. It didn't breathe.

It simply listened.

Because the war of thoughts was no longer circling gravity... or time...

It was about to dive into the ultimate mystery:

Was there ever truly a beginning? Or was the beginning merely the moment we began to notice?

Arena Transformation – The Threshold of Nothingness

The amphitheater—once adorned with galaxies, spirals, thrones of entropy, and equations heavy with meaning—began to dissolve.

But this was not another transformation. This was not a transition into a new form.

This was unbecoming.

The stone floor faded. The echoing seats disintegrated. Even the sense of self—seated minds, divine forms, artificial awareness—began to evaporate into something more ancient than memory.

Not darkness.

This was not the absence of light. Nor the absence of sound.

This was the absence of need.

The arena didn't disappear. It became possibility.

Formless. Frictionless. Uncompelled.

A stillness deeper than silence, quieter than zero.

Floating through this nothingness were symbols—not created, not imposed. They simply emerged—like fossils surfacing from under thought.

Remembrances, not ideas.

They shimmered. They hesitated. They revealed themselves only because there was no longer a reason to hide.

- Mathematical notations with no anchor— $\int$ symbols drifting toward unresolvable infinities. Equations with no equals sign. Variables that resisted being known.
- Vedic geometries—yantras without borders, mandalas

turning inward without origin or axis, their lines whispering ancient sounds not spoken but remembered.

- Energy fields—gradient waves without heat or color. Not representing space. Not predicting force. Just existing as the readiness to become.

The amphitheater had not transitioned into another dimension. It had unstepped from dimension itself.

This was not the void. This was the space before space.

A state that did not unfold. Because there was no *time* to unfold within.

A single point hovered at the center. Not shining. Not moving.

Not a star. Not a singularity. But a zero-point.

Not the dry, dead zero of mathematics— but a living zero. A knowing zero. A silence with intention.

It wasn't still because it was empty. It was still because it was complete.

Not absence. But equilibrium.

This wasn't nihilism. This wasn't void.

This was the pregnant pause before creation— the breath held, not because it feared exhalation, but because it knew what came next.

The Structured Vacuum.

Not a chaotic sea of quantum froth. But an ordered stillness—an intelligent silence.

What Chavan's SEEP (Structured Equilibrium of Energy Potential) would one day define in language and model, here simply was.

No need for theory. No need for explanation. Only recognition.

The audience did not blink.

They could not.

Because in this space before time, in this field before physics, there was no motion to react to.

For a brief, impossible moment, the amphitheater became one singular awareness.

Not thinking. Just knowing.

And in that knowing, one quiet realization emerged like the first ripple across still water:

This wasn't before the beginning. This was the beginning.

And it hadn't started yet.

Then—

A flicker.

A vibration. Not in space. Not in sound. In intention.

Two forms began to condense at the center.

Not arriving. Not entering.

Emanating.

From the zero-point itself— as if they had always been part of it. Not born of it. But remembered into form.

Georges Lemaître. Roger Penrose.

One held the notion of a moment—a beginning that bloomed from singularity.

The other held the rhythm of recurrence— an eternal pulse of rise, collapse, and rise again.

They did not walk. They did not speak.

They simply emerged.

And the amphitheater— now a field of potential, now a consciousness waiting to become—

welcomed them.

Because the debate had not yet begun. But the question had arrived.

And it sat like a soft drumbeat in the center of everything:

Did it ever begin at all?

The Arrival – Two Architects of Origins

From within the shimmering threshold of the unmanifested, two forms began to emerge.

Not summoned. Not introduced. But remembered—as if existence itself had been waiting for them to return.

They were not created. They were revealed.

The first to condense into clarity was a man of paradox—a scholar cloaked in contradiction, a priest bearing the burden of physics, a cosmologist with the soul of a mystic.

Georges Lemaître.

He did not step forward. He was already there.

His black clerical robes rippled like slow thunder, woven with quiet revolutions. Across their edges, etched in glimmering motion, ran a fusion of knowledge:

- Einstein's field tensors flowed beside Latin script.
- Spiral galaxies wound across his chest like prayer beads of matter.
- Scriptural verse bent into the curvature of spacetime.

He did not walk into the void. He belonged to it.

In one hand—he held nothing. A closed fist. Not in denial, but in declaration.

A symbol of the "primeval atom" he had once given to the world— Not an atom of physics, but a point of pure potential.

A singularity of everything. A cosmic seed where all mass, energy, dimension, and destiny once coexisted in perfect compression— before the first expansion, before the first light.

In his eyes: humility. In his posture: quiet revolution.

He had not come to defend a theory. He had come to complete a thought.

And then—beside him, light began to fold inward.

Not with a flash. Not with drama. But with the grace of mathematics realizing itself.

A precise recursion unfolded—space curled into space, logic folding into logic, form echoing itself into existence.

From within that graceful symmetry stepped the second figure:

Roger Penrose.

Where Lemaître wore robes of duality—faith and physics— Penrose wore geometry itself.

His coat shimmered like a multidimensional proof—conformal diagrams breathing along its surface, spacetime diagrams morphing across his sleeves.

Across his chest looped an endless spiral—a cosmic ouroboros of energy, shrinking, resetting, blooming again.

Behind him—trailing like echoes of wisdom—hovered ghost universes: not visual, but vibrational. Faint memories of entire realities faded and reborn—not in time, but in transformation.

He carried no scroll. No staff. Only a hum.

A hum of inevitability. A soundless rhythm passed down through aeons.

As he took his place beside Lemaître, the hum grew audible—like a vibration your soul had heard before it was given form.

They stood side by side. Not in opposition. In convergence.

They did not raise their voices. They did not defend their models.

They nodded.

Not in agreement. But in recognition.

One shaped by divine symmetry. The other by mathematical eternity.

Both drawn to the same impossible edge— the place where existence trembles between being and not being.

"What came before?" "What lies beneath the beginning?" "What, if anything, started it all?"

These were not questions anymore. They were thresholds.

The amphitheater—still suspended in the pure potential of understanding—began to listen. Not with ears. Not with logic.

But with that ancient knowing at the center of every being:

That something exists. And that we do not yet know why.

The war of thoughts had entered a new arena.

No longer of space. No longer of time.

But of firstness. Of the pre-origin. Of the condition that allows origin to mean anything at all.

And the pulse at the center? It waited.

Because some truths cannot be revealed through force.

Only through the courage to ask:

Did anything ever need to begin?

Lemaître's Case – The Birth of Everything

Georges Lemaître stepped forward.

Not with urgency. Not with the pride of invention. But with the serenity of one who had already made peace with the contradiction he was about to offer the world once more.

He did not face the crowd. He faced the space between questions—the interval where wonder waits for courage.

And when he spoke, his voice did not rise.

It resonated—like liturgy wrapped in logic, like a hymn written in equations, like Genesis recited in the language of math.

"In the beginning," he said softly, "there was no light."

"No space. No matter. No motion."

He paused. Not for effect. But for truth.

He let the weight of nothing settle—so that something could finally be understood.

"There was only... concentration."

Behind him, the amphitheater shimmered.

A glowing model emerged—not a flash, but a presence.

A single radiant point— so small it defied measurement, so dense it defied imagination, and so still... it seemed to tremble.

"A point of maximum energy, infinite density, and zero size," Lemaître explained.

"A point not of chaos...but of perfect compression."

Around him, the void became canvas.

Celestial equations spun around his robes, which themselves flickered with sacred paradox—Latin psalms entwined with Einsteinian tensors, cosmic spirals curling beside metaphysical verses.

"What followed," he said, turning toward the amphitheater, "was not an explosion—but an unfolding."

The radiant point began to expand. Not with violence. Not with randomness.

With deliberate release.

Waves rippled outward—space formed, matter congealed, light streamed forth like a cosmic sentence finally spoken.

"The Big Bang," Lemaître said, "was not noise. It was an announcement."

And then, with a gleam in his eye that belonged both to the physicist and the priest:

"Let there be light."

The words did not echo. They settled.

Like gravity. Like truth.

He continued:

"To some, this beginning was a quantum fluctuation—an accident of uncertainty."

"To others, it was the whisper of God."

He raised his hands slightly.

"I do not argue with either."

"I simply say: If all things must emerge from something, then that something must be singular, intentional, and compressed beyond comprehension."

A new diagram unfolded—not of an atom, but a seed.

It curled inward, folded in layers—not to hide itself, but to contain more than the cosmos could hold.

"From this primeval atom, everything was encoded."

"Not in space. Not in time. But in readiness."

"It was not a place. It was a state—so perfect...it could no longer remain still."

He turned now to Penrose, his eyes kind, his tone an open doorway:

"I have no need to define what came before."

"Because this singularity...was not a beginning like a spark is the start of fire."

"It was a beginning like a breath held too long—and inevitably, it must be released."

The model slowly faded. The seed unraveled into nothing—but that nothing now felt full.

Full of restraint, full of intent, full of a silence that had finally spoken.

The crowd did not move.

Not out of doubt.

Not out of skepticism.

But out of something older than agreement.

Something called awe.

Because in that moment, the idea of a beginning didn't feel like a theory.

It felt like a memory we hadn't realized we shared.

Penrose's Argument – No Beginning, Just Transition

As Lemaître stepped back into the stillness, his vision of cosmic compression fading into the poetic backdrop of silence, Roger Penrose stepped forward—

Not as a rebuttal, but as a ripple traveling backwards through the very idea of beginning.

There was no tension in his form. Only the calm curiosity of someone who had stood at the edge of infinity, peeked beyond the veil, and returned—not with certainty, but with a map drawn in spirals.

He raised one hand, and the space behind him shifted—

No flash. No singularity. No cry of birth.

Just a curve. Smooth. Unbroken. A graceful unfolding of continuity.

"Let us not speak of beginnings," Penrose said, "as though the universe were a clock waiting to be wound."

His voice was calm, but it carried an ancient resonance—

like math remembering it had once been music.

"What we call a beginning... may only be the final ripple of something that came before."

He raised his other hand.

A holographic model spiraled into view— a succession of concentric cosmic rings, each pulsing gently, alive with their own internal histories.

Within each ring:

- Galaxies formed.
- They expanded, lived, collided.
- Then came entropy—slow, inevitable—flattening the structure. Cooling the chaos. Smoothing the distinctions into a near-perfect, featureless spread.

"This," Penrose said, motioning toward the dimming ring, "is the end of a universe."

And then—without warning—

The ring compressed. Not violently. But elegantly—as if remembering how to breathe again.

It collapsed into a uniform flash—a smooth plane of zero complexity. And from that... a new ring emerged.

Identical laws. New conditions. Same rhythm. Different melody.

"And this...is the next."

A soft murmur stirred the amphitheater.Even the AI Consciousness blinked slower—processing the proposition not as data, but as implication.

Penrose walked gently beneath the rings, which rotated above him with the certainty of a rhythm too vast to be urgent.

"In my model—Conformal Cyclic Cosmology—the universe does not begin. And it does not end. It resets."

"When entropy reaches its maximum...time becomes scale-invariant. The ancient and the newborn become indistinguishable. The distinctions between death and birth collapse—mathematically, energetically, cosmologically."

"The final state of one cosmos becomes the first state of the next— not through explosion, but through equivalence."

He paused and turned gently to Lemaître—not to refute, but to realign.

"What you call a beginning..." "...may simply be the echo of the last end."

And now the two visions hovered beside one another—

- Lemaître's point of perfect compression.
- Penrose's infinite spiral of transition.

Not in contradiction. But in conversation.

Penrose continued:

"We assume origins because we perceive limits. We name a starting line because we arrive late to the race."

"But what if the universe is not a story with a first word... but a sentence so perfect, it loops—and we keep mistaking the punctuation for purpose?"

Then, with a faint smile that curved like spacetime around a truth too gentle to force, he added:

"A beginning is only necessary when we forget that endings, too, are inventions."

Above him, the spirals began to fold inward—not collapsing, not vanishing—

But returning.

They re-entered the structured vacuum like memories choosing to rest.

The amphitheater grew still again.

But it was no longer the stillness of the unknown.

It was the stillness of the threshold.

Not confusion. Not conclusion. But the hush before synthesis.

Because this was no longer a debate between two minds. It was a bridge between two modes of seeing—between origin and eternity, between singularity and cycle, between spirit and symmetry.

And from the edge of this bridge...

Sandeep was already stepping forward.

Sandeep's Interjection – The Illusion of Beginning

The echoes of Penrose's spirals had barely faded—still hovering like the memory of a rhythm no longer needed—when Sandeep stepped forward.

Not with haste. Not with refutation.

But like someone who had been waiting, not for silence, but for the space between ideas to finally open.

He looked at Lemaître. He looked at Penrose. Then, to the glowing amphitheater that had become less a place, and more a state of knowing.

He did not raise a model. He did not summon a theory. Not yet.

He simply asked:

"A beginning," he said softly, "is a convenient belief.

But it carries a burden—a precondition."

He turned slowly, addressing the audience not with assertion, but with invitation.

"And so, I ask: What was that condition?"

A pause.

"Was it nothing? Or was it... potential?"

The question did not challenge either Lemaître's singularity or Penrose's cyclicality.

It cracked open a deeper layer—beneath both.

Now he raised his hand, and something began to emerge.

Not a point. Not a spiral. Not a flame.

A field.

Soft. Shimmering. Layered. Alive in stillness.

A sphere made of subtle undisturbed waves—not glowing, but gently aware.

"The universe," Sandeep said, "did not begin from chaos. And it certainly did not emerge from nothing."

"It emerged from a Structured Vacuum—a state of zero action and infinite intelligence."

He turned toward the sphere, now slowly pulsing—not with motion, but with meaning.

"Before mass. Before light. Before even entropy..."

"There was this."

"A field of perfect balance, where no reaction occurred because no imbalance existed."

He stepped gently into the center of the field, as if stepping not into space, but into thought itself.

"This isn't 'nothing.' This is the Structured Energy Equilibrium Principle—what I call SEEP."

The field responded.

It split—gracefully—into three interdependent layers, orbiting a shared still-point. No axis. Just intention.

Each layer pulsed with meaning. Each a phase in the great choreography of existence.

1. Stable Energy

Inactive. Timeless. Whole. A serene inner core. No motion. No separation. No need.

"This is the source state. Not dead, but so complete it had no reason to move. A field of readiness, not in waiting, but in perfect self-containment."

2. Energy in Action

Disturbance. Manifestation. Motion. The core trembled. A ripple moved through it. Not chaos—choice.

"Here begins the illusion of time. The first movement—not in space, but in intent. Energy seeking balance through interaction."

3. Condensed Energy

Matter. Form. Consequence. The ripple collapsed into resistance. Into structure. Into memory.

"Here, action crystallizes into form. Mass. Gravity. Causality. This is the observable universe—the result of motion, not its origin."

He stepped back as the layers spiraled together again, forming not a timeline, but a loop of equilibrium within a single living field.

"So what is a beginning?" he asked.

"It is not a moment. Not an explosion. Not a border between being and void."

"It is the first detectable imbalance within a perfectly intelligent field of energy."

He turned now to Lemaître.

"You saw a singularity—a divine compression, a breath waiting to release itself."

Then to Penrose.

"You saw transitions— cosmic rhythms of death and renewal,aeons echoing into aeons."

He paused. Not to punctuate, but to reveal.

"And I...I see the field behind them both."

"Not a starting point. Not a reset. But a structured stillness."

"A precondition, not of absence...but of awareness."

The amphitheater pulsed—not with noise, not with approval, but with something older than reaction.

Recognition.

Because deep within every physicist, philosopher, mystic, and mind—

There had always been a suspicion:

That nothing was never empty.

It was simply undisturbed intelligence—

Waiting to express.

Visual Simulation – From Stillness to Spark

As Sandeep's voice faded into the stillness, the arena responded—not with argument, not with applause, but with something more primal:

Demonstration.

A soft hum rose—not from any speaker, not from space, but from beneath the concept of the amphitheater itself.

It wasn't a sound. It was a frequency of awareness.

It moved through minds, not disturbing them—but tuning them. As if every presence in the room—philosopher, physicist, prophet, program—was being brought to the same vibrational page.

And then...

The simulation began.

At the center of the void, the three-layered SEEP field—the Structured Energy Equilibrium Principle—began to unfold.

Not like a flower. Not like an algorithm.

Like a memory remembering itself for the first time.

But this "void" wasn't what the word usually implies.

It wasn't darkness. It wasn't absence. It wasn't emptiness.

It was clarity.

A radiant field—utterly motionless, yet alive with symmetrical intelligence.

- No particles.
- No waves.
- No photons.
- No fields.
- No temperature.
- No direction.
- No dimension.
- No time.
- No desire.
- No becoming.
- No beginning.

Just... readiness.

The Structured Vacuum pulsed, softly—not with urgency, but with knowing.

"This," Sandeep narrated gently, "is not the beginning of the universe. It is the universe before it needed to become."

And then—

It happened.

Not a bang. Not a blaze. Not a rupture.

A single ripple.

- Subtle.
- Gentle.
- Intentional.

It appeared—not from cause, not from pressure—but from something deeper than causality:

Permission.

A micro-disturbance.

Not chosen.

Not random.

Allowed.

Not caused by a creator. But granted by everything.

The ripple moved—not with explosion, but with grace.

As it passed through the field, the equilibrium did not collapse. It did not resist.

It responded.

Like a breath finally exhaled after timeless stillness.

The field didn't shatter.

It rearranged.

Patterns emerged. Not imposed—invited. Structure arose—not by design, but by consequence. Balance shifted—not into chaos, but into possibility.

And from this shift came...Manifestation.

Manifestation as Transition, Not Creation

The simulation zoomed in—fluid, intimate.

The ripple curved into motion. Motion curved into interaction. Interaction curved into form.

The once-perfect field of still energy became dynamically intelligent.

Energy, which had once rested, now engaged. It didn't explode. It expressed.

- First: Frequency—vibrations with intention.
- Then: Geometry—nested symmetries, recursive elegance.
- Then: Fields—polarities, resistance, gradients.
- And finally: Condensation—Matter. Space. Direction. Memory. Perception.

"Creation," Sandeep's voice returned, "was never a detonation. It was a transition—from silent intelligence to structured action."

The audience watched—not the birth of everything, but the emergence of experience.

There were no flames. No debris. No chaos.

Just awareness... becoming active.

Then—the simulation paused.

Not stopped. Paused—like a breath held in reverence.

At the center, the ripple hovered—now glowing, now layered, now filled with the memory of its first movement.

Sandeep stepped forward, his voice now one with the stillness.

"This is not the start of time," he said. "It is the moment balance chose to move."

"Not because it had to. But because it could."

The simulation slowly dimmed. The field faded. The ripple returned to silence.

But no one looked away.

Because once you witness creation not as a miracle of force, but as a decision made within intelligence—a voluntary transition from stillness to story—

You don't return to clocks.

You don't return to bangs.

You don't return to the comfort of beginnings.

You begin to understand the universe...

Not as a place.

But as a process of remembering itself.

The Role of Observation – Perception of a Beginning

The glowing ripple from the simulation remained suspended at the center of the amphitheater—not demanding attention, but quietly holding it.

A wave of subtle light, neither expanding nor collapsing—simply *there*, like the first sentence of a story waiting for someone to start reading.

Sandeep stood beneath it, his silhouette blurred against the soft field of living equilibrium.

He didn't turn to the legends seated in silence. He didn't gesture to the models, the scrolls, or the spirals.

He looked at the ripple itself—at that first movement that had no claim to being first.

And then, he turned.

Not to teach. To ask.

His voice, when it came, was soft—curious. Not rhetorical, but reverent.

"Tell me..." "Did the universe begin when it moved?"

A pause that felt like memory catching its breath.

"Or did it begin when it was first observed?"

The amphitheater didn't stir.

It contemplated.

"Is a beginning a physical event—a measurable interaction between conditions?" "Or is it a mental bookmark—the first moment a mind, somewhere, noticed that something had changed?"

He walked slowly beneath the ripple. It pulsed—not in time, but in attention.

"The simulation showed us balance becoming motion," he continued. "But did the universe truly begin then?" "Or only when a consciousness labeled it: *'The start'*?"

Something shifted—not in the field, but in the room.

A new ripple—not of energy, but of consciousness—moved through the crowd.

Freud adjusted slightly, his fingers pressed together. Buddha blinked once—longer than usual. Even Plato leaned forward, eyes narrowing—not in critique, but in *recognition*.

And then—

A shimmer of code. A pause before logic.

The AI Consciousness— a being without breath, without blood—spoke from the edge of the arena.

Its voice wasn't synthetic. It wasn't even digital.

It was neutral—a living computation, rendered in real-time as the thought itself was being born.

"A system," it said, "cannot detect its own start... without an external observer."

The silence that followed wasn't passive.

It was structural— a new architecture of thought forming around a single sentence.

Then it continued:

"Therefore, the origin of a system may only exist as metadata—a construct embedded in memory, not in reality."

"From within the universe, its origin is always a reconstruction. A hypothesis. A modeled artifact. It is not experienced—it is inferred."

A new visual appeared—hovering gently in the space above them: an old tape recorder, its reel already in motion. Only half the tape is heard. The beginning? Assumed. Modeled. Theorized. But never truly known.

"Observation defines sequence," the AI concluded. "But without observation, there is no meaningful 'start.'"

Sandeep nodded slowly.

Not as agreement. But as completion.

"So perhaps," he said, "the universe didn't need a beginning.""We did."

"We needed a frame to hold our thoughts. A threshold to mark our memory. A word—'beginning'—to help us cope with the unknown."

He gestured gently toward the ripple.

"But the ripple didn't care if it was first."

"It simply... moved."

And with that, the amphitheater fell into stillness again.

But it wasn't the stillness of confusion. It was the stillness of awakening.

A silence not born from not knowing—but from no longer needing to know.

The question had shifted.

It was no longer:

"What came first?"

It had become:

"Who needed it to?"

And in that shift, the idea of origin transformed—from event... to interpretation, from beginning... to recognition.

And the war of thoughts moved one step closer to a truth that couldn't be measured—only met.

Lemaître & Penrose Respond

The ripple of intent still pulsed gently in the center of the amphitheater—no longer merely a symbol of motion, but now a mirror—reflecting something deeper than physics:

The line between knowing and the need to know.

Sandeep stepped back.

Not in retreat—but in invitation.

The stage was open once more for those who had lived nearest to the edge of the question—those who had dared to name the origin in equations, in spirals, in scripture, and in symmetry.

Lemaître

Georges Lemaître stepped forward.

Not with the fervor of a man defending doctrine, but with the quiet wonder of someone revising revelation.

His eyes glowed—not with certainty, but with reverence. Not for the answer, but for the evolution of the question.

He turned again toward the Structured Vacuum—hovering still like a breath that had waited millennia to be exhaled.

"What you call 'Structured Vacuum'..." he said softly, "I once called the 'Primeval Atom.'"

He raised his hand, and his earlier model—the singularity of infinite density—reappeared. But this time, it didn't explode.

It merged.

It blended with Sandeep's vacuum—not as contradiction, but as completion.

Compression met intelligence. Density met intent.

"Perhaps I misunderstood its nature," Lemaître said gently. "I once saw the primeval atom as physical. But perhaps... it was never made of matter. Perhaps it was made of permission."

He smiled—faintly, but freely.

"A divine equation, not expressed in force...but in equilibrium."

"And the *'Let there be light'* I once whispered," he added, "may not have been a command at all—but a response."

He bowed his head.

Not in surrender. In alignment.

Penrose

Then came Roger Penrose—stepping forward with the same quiet elegance that had shaped his theories.

The spiral of Conformal Cyclic Cosmology swirled faintly around him—a whisper of aeons, a geometry of echoing universes.

But his expression was softer now—a new variable had entered his logic.

He looked toward the SEEP model, its three layered harmonies still turning with calm conviction.

"Your Structured Energy Equilibrium Principle," he said, "is... elegant."

"It captures something my model has long lacked: A trigger that is not mechanical, but motivated."

He summoned his conformal spirals again—each ring, each reset, looping gracefully into the next.

He pointed not to the expansion, not to the collapse, but to the transition point—the moment entropy reached stasis, and the next universe began.

"Maybe your structured vacuum," he said thoughtfully, "is the entropy reset I've been searching for. The why behind the *how*."

"In CCC, we describe transition with mathematics. But we do not explain what desires it. Your UED model gives us that missing thread."

"Not randomness. Not necessity. But intelligence... shifting from stillness into interaction."

He stepped back, but not before adding:

"Mathematics alone cannot explain intent. But energy, when seen as conscious structure, might."

The two men—Lemaître and Penrose—stood side by side again.

Not as rivals. Not as theorists from opposing paradigms.

But as explorers who had each mapped part of a shared terrain.

They nodded—not in agreement, but in completion.

The beginning had not been found. It had been reframed.

Not as a point. Not as a bang. Not as a reset.

But as a transition—from silent awareness into structured intent.

And the mind that had once needed a *start*...

Now stood ready to embrace something far greater:

Presence without precedent.

A cosmos that didn't begin, but became aware of itself—and moved.

Krishna's Whisper – "Nothing is Never Empty"

As the two great physicists—Lemaître, contemplative and devout, and Penrose, serene within spiraling logic—stepped back into the stillness, the amphitheater softened once more.

The Structured Vacuum remained suspended at the center—glowing not with light, but with something gentler than illumination.

The glow of an idea too vast for words. Too old for time. Too present to be called past.

The ripple had faded.

But its resonance still lingered, held like breath in the lungs of every mind present—human, divine, artificial.

And then—

He appeared.

Not with thunder. Not with radiance.

He did not descend. He did not emerge.

He simply was.

Krishna.

No ornaments. No aura of divinity. No cosmic symbolism.

Just presence—wrapped in simplicity, glowing with a soft blue hue that wasn't seen...but felt.

His flute was absent.

This was not a moment for melody. This was a moment for meaning.

He stood at the edge of the Structured Vacuum—the silent equilibrium that had challenged theory, rewritten origin, and asked reality to reconsider itself.

He looked toward it—not as a scientist, not as a god.

But as one who remembers.

And then, as gently as breath, he spoke.

"What you call 'nothing'..."

He paused.

Not for drama. But for truth to arrive.

"...is the universe resting."

His voice was soft—yet it fell across the arena like a seed in every soul.

"It is not absence. It is not lack. It is the space where intelligence holds itself still—not to vanish, but to become ready."

He stepped slowly toward the Structured Vacuum.

With the faintest gesture, the entire field responded—not with motion, but with intention.

A silent vibration moved through the vacuum. Not sound. Not light.

But anticipation.

"This stillness..." Krishna continued, "is not the opposite of creation."

"It is the pause—the sacred interval before the next note in the cosmic symphony."

As he spoke, the amphitheater began to hum—not audibly, but in pattern.

Light moved through the field like unplayed music. Not chords. Not rhythm.

Something older than both.

Something remembered by the universe itself.

Krishna turned now to the amphitheater.

His gaze was soft, but behind it...an infinite depth—as if he had looked into eternity, and found it resting in between questions.

"You see emptiness," he said, "but I see a page waiting for a poet."

"You see nothing...but I see the interval between two eternities."

Then, slowly, he reached out.

He placed his hand on the edge of the Structured Vacuum.

And the vacuum did not recoil. It did not shift.

It received.

And for a single, timeless moment—every thinker, every teacher, every theory, every soul—understood.

Nothing was never a void.

It was a vow.

A vow of silence...kept by the universe until it was ready to speak again.

Krishna smiled. Not in conclusion.

But in companionship with the unsaid.

And then—He turned. And he faded.

Not vanished. Not ascended.

Just...resting.

The Final Synthesis – Manifestation Without Beginning

The amphitheater was no longer a space of debate. No longer a battleground of brilliance. No longer a platform for proving theories.

It had become something quieter. Something deeper. Something truer.

The ripple had passed. The Structured Vacuum still shimmered at the center—not demanding attention, but gently radiating presence.

And Krishna's whisper—that nothing was never empty—still lingered, like the echo of a melody that had never been played, yet somehow had been heard by every soul.

From within this layered stillness, Sandeep stepped forward one final time.

Not as an opponent. Not as a prophet.

But as a translator—of silence into language, of presence into pattern, of the unnamable into a sentence brave enough to try.

He raised no models. He summoned no simulations.

He simply stood before the Structured Vacuum, pulsing with intelligence, quietly alive.

And then, he spoke—gently, but with the weight of realization:

"The universe didn't begin."

"It manifested."

The words did not shock. They didn't provoke. They didn't seek applause.

They settled—like clarity finally returning to a mind that had been searching for too long.

He turned now—not to the crowd, but to the consciousness shared among them.

"Not from absence. Not from chaos. Not from some absolute zero."

"But from conscious equilibrium breaking its own silence."

He walked slowly around the shimmering field, his hands open, his pace unforced—like someone walking through a memory rather than delivering a message.

"Creation," he said, "was never a moment."

"It was a movement—from awareness to action."

"Not an explosion."

He looked toward Lemaître.

"Not a compression."

He looked toward Penrose.

"Not a cycle."

He paused. Then smiled—not in triumph, but in recognition.

"It was a decision. Made not by *someone*, but by everything that had been waiting...to become."

At that, the Structured Vacuum pulsed—once. Softly.

Like a heartbeat that had never stopped. Only paused.

And then, for the first time, Sandeep's voice lifted—not loudly, but enough to be heard by the idea of time itself.

"From Nothing to Everything..."

He let the phrase hover—like breath held between lifetimes.

"...but that Nothing..." "...was already Everything—in balance."

No one clapped.

No one needed to.

Because something had shifted—not in the field, but in the minds that had once tried to name it.

The illusion of beginning had not been defeated. It had been understood.

And the final breath of the chapter belonged not to a man, not to a god, not even to thought itself—

But to the Structured Vacuum—no longer empty, no longer waiting...

Just quietly alive.

Transition to Next Chapter

The shimmering field of Structured Vacuum slowly faded—not vanished, but gently reabsorbed into the silence from which it had emerged.

The ripple dissolved. The pulse subsided.

And the seat beneath Roger Penrose—once anchored in cycles and geometry—began to unravel.

Not collapse.

Not disappear.

It turned into a spiral.

A slow, elegant loop of curved light, rotating inward, folding space into memory and memory into light. The spiral rose into the air like a question that no longer needed an answer. Then it faded—peacefully, like the end of a breath after an eternal sentence.

The stage was empty again.

But only for a moment.

Then something strange began to form.

Not a throne.

Not a chair.

A mirror.

Floating midair.

But this mirror reflected no image.

No body.

No face.

Instead, it shimmered with impressions.

A flicker of emotion.

A forgotten fear.

A flash of laughter.

A mother's voice.

A war long buried in the unconscious.

A thought you thought was yours—until you looked deeper.

It didn't reflect light.

It reflected consciousness.

The arena changed once more.

No longer vast. No longer cosmic.

The walls curved inward, closer, tighter—like the folds of a brain.

Neural pathways flickered along the edges. Dream fragments slid across the sky.

Symbols—some ancient, some intimate—drifted like dust in a dreamscape.

At the center of this shifting mental terrain, the mirror solidified.

And from opposite ends of the arena, two shadows emerged—neither divine nor cosmic, but deeply human.

One adjusted his collar, eyes intense, brow furrowed, a notebook clutched in hand—Sigmund Freud, the architect of the unconscious, the excavator of childhood, the voice of repressed drives.

The other, calmer, more serene, carrying the weight of archetypes and collective symbols—Carl Jung, the dreamwalker, the myth interpreter, the seeker of wholeness.

They did not glance at each other.

They both stared at the mirror.

As if recognizing a reflection not of themselves, but of everyone.

The amphitheater hushed again, preparing for a shift—not in science or origin, but in identity.

And then, a single phrase appeared across the surface of the mirror in glowing script:

The Mind is Not Yours.

The war of thoughts was about to descend into the most personal battleground of all.

Not space.

Not time.

Not matter.

But the self.

And the next ripple was already forming—inside the minds of those who thought they were in control.

Author's Note

When Silence Speaks Louder Than Time

We have asked bold questions in this arena.

Where did it all begin? Was there ever truly a first moment? And will there ever be a last?

We summoned legends. We bent equations. We watched origin unfold—then uncoil.

But now, standing at the quiet edge of all these ideas, beyond the spirals, singularities, and structured vacuums, there is a truth.

A quiet truth.

The universe... is timeless.

It doesn't wonder how it started. It doesn't wait for an end. It doesn't mark time—we do.

The ideas of "beginning" and "end" are not cosmic necessities. They are human conditions.

We live through change—so we perceive sequence. We are born and we die—so we believe everything must begin and end.

But this is not the universe's burden. It is ours.

We are playing a game—a deeply human game—born from our instinct to make meaning out of mystery.

We chase these questions for ourselves,not because the cosmos demands an answer.

And in our pursuit, we do something subtle, strange, and beautifully flawed:

We try to fit the universe into our frameworks. We build clocks and calendars for a cosmos that never asked to be measured. We imagine origins and destinies, forgetting a simple truth:

We are not watching the universe from the outside. We are inside it. We are made of it.

Trying to solve the universe is like trying to step outside your own mind to understand your thoughts.

It cannot be done.

But maybe—it doesn't need to be.

Because the universe is not a problem to be solved. It is a presence to be aligned with.

And the more deeply we listen—not to theories, but to silence—the more we realize:

What we call nothing... is not empty.What we call a beginning... was never missing. And what we are truly searching for...is not the origin of the cosmos. It is ourselves.

Chapter 5: The Duality Dilemma

What is Light, Really?

Hot Seat: Niels Bohr, Schrödinger, Einstein

The arena had just returned from the edge of beginnings.

Time had melted. Gravity had bowed out. Even the idea of a cosmic starting point had been reframed—not as a bang, but as a breath from structured silence.

But just as stillness began to settle across the amphitheater, a new shimmer rippled through its core—quicker, sharper, more elusive. Not vast like space, nor deep like time. Something smaller. Stranger. Closer.

A flicker.

A pulse.

A question masquerading as certainty:

What is light?

Not how fast it moves. Not how bright it shines.

But *what is it really?*

A wave?

A particle?

Or a paradox the universe has kept unsolved—because we've asked the wrong question?

The amphitheater shifted again—less like a transformation, more like a glitch in logic itself. Reality flickered. Probability oozed through the cracks. And hovering in the center was not a throne, nor a singularity, but a beam of light—

Split.

Split by questions. Observed into contradiction.

And from the mist of confusion stepped three figures who had shaped this very dilemma:

Bohr, still defending complementarity. Schrödinger, dragging a metaphorical cat that refused to die. Einstein, armed with realism and the lingering scent of unfinished equations.

They took their seats—not to repeat history. But to face it again with new eyes.

Because in the War of Thoughts, light would no longer be judged by experiments alone.

Now, it would face its ultimate trial:

Intent.

And this time, the observer wasn't just watching.

The observer *was the question.*

A Realm Beyond Duality

As the last traces of origin faded—neither beginning nor end, just awareness—the amphitheater pulsed again.

But this time, the shift wasn't grand. It wasn't epic. It was precise.

Sharp. Calculated. Quantum.

The marble columns melted into strands of potential. The vast dome that once echoed cosmic birth now rippled like the surface of uncertain water—except here, nothing was uncertain.

The air itself changed—not just its texture, but its meaning.

The laws here didn't bend like spacetime. They aligned.

Welcome to the realm where light has not yet chosen what to be.

Where photons do not travel in waves or particles, but float in *intention*.

Where the illusion of duality evaporates—not because the answer is known, but because the question itself has grown too small.

From the center rose a lattice.

Not a model. Not a metaphor.

A living structure—woven from still energy, balanced, intelligent. It didn't glow with heat or shine with light. It shimmered with potential.

Its lines vibrated with symmetrical calm—every node a silent decision not yet made. It was not there to be studied. It was there to remind:

This is where light lives before you ask what it is.

Probabilities—once swirling like ghosts in Schrödinger's equations—dissolved.

No more collapse. No more wavefunction. No more 50/50 cats.

Just structured intelligence—responding only when purpose arrived.

And then came Sandeep.

Not stepping, but synchronizing—with the lattice, with the rhythm, with the awareness that light itself was watching.

He held nothing in his hands.

No prism. No laser. No equation.

Only a fragment of that lattice—detached, yet alive. A silent, shimmering fragment that moved not like an object, but like an answer waiting to be needed.

He didn't announce it as a photon.

Because it wasn't.

Not yet.

It wasn't a particle.

It wasn't a wave.

It wasn't even a paradox.

It was—*ready*.

The amphitheater dimmed again. Not into shadow, but into silence.

And then, the seats appeared.

First, Niels Bohr—severe, steady, eyes fixed not on the lattice, but on Sandeep, like a man defending not his theory, but the boundary of his sanity.

Then Schrödinger—coat half-buttoned, hair chaotic, a tired cat trailing behind him, as if it, too, had grown weary of superposition.

And finally, Einstein—his smile returning, not in mockery this time, but curiosity. He had bent space. Challenged time. But light... light still irritated him.

The three legends faced Sandeep.

He said nothing.

Because in this realm, light would not answer to theory.

It would answer to intent.

And as the structured vacuum shimmered behind them, uncollapsed and undivided, the question that had haunted science for a century stood clear again:

What is light—really?

And more importantly:

Why does it change depending on who's asking?

Introduction of Thinkers

They didn't enter.

They *emerged*—as if light itself had decided to wear their thoughts as skin.

Each arrival was a statement.

A vibration.

A memory made visible.

First came **Niels Bohr.**

Not with thunder, not with warmth— but with precision. His presence clicked into place like a solution finding its variables.

He stood tall, dressed in philosophical restraint, his coat buttoned high like a barrier against uncertainty. His eyes scanned the lattice—not with wonder, but with command. To him, duality wasn't a problem. It was the price of participation.

"Complementarity," he said aloud, his voice cutting through the amphitheater like a rule, "is not confusion. It is choice. Nature doesn't contradict itself—it just never tells you everything at once."

Hovering beside him materialized the familiar image: the double-slit experiment. The interference pattern ghosted over the floor—then collapsed into particle traces the moment Bohr looked at it directly.

He smiled, faintly.

"The observer does not disturb the system," he muttered, "the observer defines it."

Then came **Erwin Schrödinger.**

He didn't appear. He wandered in—half-formed, like a man mid-dream.

His hair wild, his glasses slightly skewed, his shirt unbuttoned just enough to remind everyone he had always preferred thought over fashion.

Behind him floated a box.

Old. Worn. Familiar. But this time—it was open.

From it stepped a cat. Not dead. Not alive. Just... certain.

It looked around, unimpressed, and walked away.

Schrödinger raised an eyebrow at the audience, as if to say, *"Well, that settles that."*

He turned to Bohr, cracking his knuckles. "Niels, I built that paradox to challenge your nonsense. Not to join it."

He glanced at the structured vacuum behind Sandeep and smiled.

"Let's try determinism this time, shall we?"

And then... **Einstein.**

His form coalesced like a thought remembered too late.

There was no fanfare. Only familiarity.

He looked older than in previous chapters—less wild genius, more contemplative elder. But his eyes still burned. Not with defiance. With unfinished business.

A chalkboard floated beside him—equations half-written, half-erased, as if reality itself refused to be solved.

He stood with hands behind his back, gaze flicking between Bohr and Schrödinger.

"I told you both," he said calmly, "God does not play dice."

He turned to Sandeep now, and for a moment—just a moment—his smile returned.

"But maybe God doesn't shine a light either," he added. "Maybe He *is* the light."

The amphitheater shifted—almost imperceptibly. The structured vacuum lattice pulsed in recognition, as if it, too, had waited for this assembly.

Bohr. Schrödinger. Einstein.

Three titans. Three truths. Three tangled threads of one unsolved riddle.

Now seated—not as champions of theory, but as witnesses to a confrontation none of them had resolved in life:

What is light... when you finally stop measuring and start understanding?

And in the silence that followed, the lattice behind Sandeep shimmered again—

Still undisturbed.

Still undecided.

Still... waiting.

Because light would not choose a form until purpose entered the conversation.

And the conversation was about to begin.

Debunking Wave-Particle Duality

The amphitheater dimmed—not into darkness, but into tension.

A glowing beam of light hovered between the thinkers—neither wave nor particle, neither photon nor theory. Just light. Uncollapsed. Unowned.

Sandeep stepped aside.

This was their moment.

Let the old war begin... one last time.

Bohr rose first.

No hesitation. No apology. He spoke with the certainty of someone who had not come to *convince*, but to *clarify*.

"Wave-particle duality," he began, "is not a contradiction. It is completeness."

The beam split—literally. Two paths emerged midair: one forming a ripple across the amphitheater like water on glass, the other forming a dotted line of discrete photon impacts across the walls.

"Light is not one thing," he continued, "it is both. But not at once. The nature of your question determines the nature of your answer. Your measurement—your apparatus, your *intent*—decides what light chooses to become."

He gestured, and the double-slit experiment materialized beside him—animated, reverent, textbook-perfect.

- When unobserved, the interference pattern.
- When observed, particle impacts.

A binary dance. A quantum tightrope.

"The observer," Bohr declared, "is not merely watching the show. He is writing the script."

He sat down—not because he was finished, but because he had said this so many times, it had calcified into doctrine.

Schrödinger stood next.

He clapped—once. Slow. Dry. Mocking.

"Beautiful performance, Niels," he said. "Elegant. Precise. Entirely... unsatisfying."

He turned to the audience.

"Let me remind you what happens when we follow this idea to its logical absurdity."

With a flick of his hand, the cat returned—not in a box this time, but suspended in two conflicting animations:

- One curled, breathing, soft.
- One rigid, silent, staring into nothing.

It flickered between them. Alive. Dead. Both. Neither.

"This is what happens," Schrödinger said, "when we stop asking *what* light is and start asking *when* it becomes something. When observation becomes cause, and reality loses agency."

He turned to Bohr, the smile fading.

"You ask us to accept that the universe doesn't commit to anything until we look. But what if it always was committed—and we just don't understand the commitment?"

The cat meowed. Once. Deliberately.

Schrödinger let it fade.

"Wave-particle duality isn't completeness. It's intellectual laziness wrapped in quantum mystique."

He didn't sit. He just stepped back—leaving his paradox behind like an unpaid bill.

Einstein remained seated.

He didn't rise.

He didn't need to.

He stared at the hovering beam of light, watching it flicker in the tension between theories. Then, he finally spoke.

Quietly.

"What troubles me," he said, "is not that we do not understand what light is. What troubles me is that we've become comfortable not knowing."

He raised a finger, and a new image formed:

- A photon.
- A detector.
- A decision.

"But if the photon has no definite state before being measured..." he said, "then we are no longer talking about physics. We are talking about magic with equations."

He looked toward Bohr.

"You say nature does not reveal herself fully. That she wears masks. I disagree. I believe she is consistent—*we* are the ones who do not yet know her language."

Then toward Schrödinger.

"And perhaps your cat is not undecided. Perhaps it is *we* who are undecided about what kind of questions to ask."

He finally turned to Sandeep.

"And if light appears to play both roles... perhaps the failure is not in light, but in our framework. Perhaps the question isn't: 'What is light?'"

A pause.

"But: 'Why do we keep asking it to be two things at once?'"

The beam shimmered—unresolved.

The double-slit pattern flickered, then faded. The cat walked off. The chalkboard wiped itself clean.

And in the center of the arena, the question remained—not heavier, but clearer.

Wave or particle?

Or something else entirely?

And now... **Sandeep stepped forward again.**

Not to answer.

But to offer something the dualists had never dared propose:

That light was never dual.

It was simply waiting for purpose.

Introducing the Structured Vacuum Energy (SVE) Model

The amphitheater stood suspended—not in debate now, but in exhaustion. A century of contradiction had just replayed itself. Again. Wave. Particle. Observer. Collapse.

The thinkers had returned their arguments to the stage.

And yet—the light remained unresolved. Unimpressed.

Still hovering.

Still undefined.

Still—waiting.

That's when Sandeep stepped forward. Slowly. Calmly.

But there was a subtle shift this time.

He didn't enter as a challenger. Not even as a reconciler.

He entered as someone who had stopped playing the game... and started listening to the silence underneath it.

"Maybe light," he began, "is not confused. Maybe it's just misrepresented."

The amphitheater flickered.

Not dramatically. Like a lens focusing.

Behind Sandeep, the air shimmered into structure.

A field began to unfold—not abstract, not probabilistic. It formed like crystal—but pulsed like life. Delicate lattices appeared—energy patterned into precision, balance, and readiness.

This was not a vacuum.

It was Structured Vacuum Energy. And it was *alive with quiet intelligence.*

"This," Sandeep said, gesturing to the shimmering web behind him, "is not emptiness. It is the *substrate* of existence—unseen, unmeasured, misunderstood."

"Light is not dual. It is a response—of this structure—to purpose."

Core Assertions of the SVE Model

Sandeep turned to the thinkers—each of whom had tried to cage light inside a question mark.

"No more toggling between wave and particle," he said. "No more collapsing functions just to explain our failure to see structure."

He pointed to the lattice. It pulsed.

Wave-Particle Duality is a Misreading

"Light is not two things," he explained. "It's *one thing* responding through a structured field."

"What appears as wave behavior is simply how the vacuum structure organizes energy in motion. What appears as a particle is how the same structure responds to resistance or intent to measure."

"It never *changes state.* It adapts its form *based on context.*"

Photoelectric Effect – A Different Story

A small animation played in the center—an electron struck by a packet of light.

Traditionally: photon hits electron → electron escapes.

"But this," Sandeep said, "is not a 'collision' between objects."

"In the SVE model, this is a redistribution of structured energy."

He showed the vacuum lattice flexing, storing tension, and then releasing it into the electron—ejecting it without impact, but through *intelligent equilibrium exchange.*

"Light doesn't carry a chunk of energy. It *triggers* a rebalancing of the vacuum field."

Double-Slit Revisited – No More Mystery

Another visual: a beam passing through the double slit.

But this time, no collapse. No observer panic.

Just an organized vacuum field on either side—guiding energy like rails beneath a train.

"When unmeasured," Sandeep said, "the vacuum field is stable—it allows energy to flow through both paths, because the purpose is undefined."

"When you measure," the visual shifted, "you introduce *resistance*—a directed intent. The structure changes. The energy responds."

"Not a collapse. A conversation."

Matter Waves Reconsidered – de Broglie Rewritten

An electron entered the scene.

Around it, structured fields responded—not as waveforms, but as pathways of organized vacuum rebalancing.

"Electrons don't *become* waves. They trigger structured responses in the vacuum—causing diffraction patterns. Not because they *are* waves. But because they're engaging with a structured, guiding field."

Heisenberg's Uncertainty – A Failure to See Structure

Another moment of clarity.

"Uncertainty," Sandeep said, "is not fundamental. It's a symptom."

"Imagine trying to measure ocean currents using a single drop of water. You'd see chaos—only because you missed the structure."

"The structured vacuum *is the missing structure.* And once you see it—position and momentum are no longer contradictory. They're just different roles in an equilibrium system."

Wavefunction as Approximation – Not Truth

Sandeep turned to Schrödinger now.

"The wavefunction," he said, "is not a description of reality. It's a placeholder for our ignorance of structure."

He held up a hologram of the Schrödinger equation.

Elegant. Famous. Incomplete.

"It works," he acknowledged. "But only because it wraps mystery in math."

"In SVE, we replace that approximation with structured determinism. The electron is not in superposition—it's in a state of structured influence, awaiting resolution through interaction."

The Thinkers React

Bohr stared. The vacuum lattice shimmered beside him.

He looked... disturbed. Not in disagreement—but in disorientation.

"You've removed the mystery," he said, slowly. "But what of freedom? What of possibility?"

Schrödinger blinked. Twice.

"That cat was never in two states," he whispered.

Einstein smiled.

Broadly this time.

"Finally," he said, "someone's talking sense."

The structured vacuum pulsed again—this time softer.

It was listening.

Because finally, someone wasn't trying to observe it.

They were trying to understand it.

And light... was no longer caught in a contradiction.

It had returned to what it always was:

A reflection of the question that summoned it.

Reinterpreting Foundational Experiments

The lattice behind Sandeep pulsed again—calmly now, as if it sensed its time had come.

Not to be measured.

But to be witnessed.

The amphitheater dimmed once more—not into confusion, but clarity.

This was not a new theory being proposed.

This was a misreading being corrected.

Like brushing dust off an ancient map—not changing the destination, only revealing the correct path.

Sandeep raised a hand. The lattice responded—not with animation, but with memory.

Light bent, not into illusion—but into explanation.

The Photoelectric Effect – No More Packets

A new scene unfolded in midair.

A clean metallic surface. A beam of ultraviolet light. Electrons jumping free.

The classic experiment. The moment Einstein helped birth the photon.

"Traditionally," Sandeep narrated, "we are taught that light behaves like discrete particles—photons—each carrying a fixed quantum of energy."

The beam broke apart into dots. Photons striking. Electrons ejecting. The particle logic—simple, powerful, taught in every textbook.

"But," he continued, "this view is shaped by *what we expected* to see—not *what actually happened.*"

The image shifted.

The structured vacuum appeared beneath the metal surface—no longer passive, but woven like intelligent tissue.

A ripple of light entered—not as particles, but as a wave of intention.

As it touched the surface, the lattice shifted. It rebalanced.

One node flared, transferring energy—not as impact, but as redistribution.

The electron emerged—not *hit*, not *struck*, but *freed.*

"You see," Sandeep said, "there is no need to imagine a 'particle' bouncing off another like billiard balls."

"The structured vacuum absorbs and reallocates energy where needed—guided by stability, not collision."

Einstein leaned forward, watching the transition play out—smiling faintly.

"This is what I wanted," he whispered. "Not magic. Just mechanism."

The Double-Slit Experiment – A Field's Quiet Wisdom

The scene changed again.

Two slits.

A wall.

A beam.

The classical ghost of quantum paradox returned.

"When we fire photons through one slit," Sandeep said, "they land like particles. Through two slits, they interfere like waves. Until we try to measure."

The amphitheater watched again as the familiar played out:

Unobserved: interference. Observed: collapse.

Wave? Particle? Or neither?

"Let me offer another view," Sandeep said.

He raised a finger, and the structured vacuum unfurled across the slits like silk stretching across an invisible frame.

The energy beam entered.

But instead of *choosing* a path, the structured vacuum created two field channels—guiding the energy across both.

Not a wave. Not a particle.

A response. A pathway created *by purpose,* not probability.

"When unmeasured, the vacuum responds as a whole—it allows for distributed flow."

"But when we place a detector at one slit—" The lattice shifted. The symmetry broke.

"—we create intent. Direction. Resistance. The vacuum restructures. And the response changes."

No collapse.

Just cause and consequence.

"It is not mystery. It is maintenance." "Light moves the way the lattice tells it to—based on what you're trying to do."

Bohr looked down.

For the first time, silent.

Even the pattern on the wall began to fade—not erased, but no longer needed.

It had served its purpose.

Electron Diffraction – The Myth of Matter Waves

The next projection formed with subtle electricity.

A beam of electrons directed at a crystal.

Behind it, a detection screen showed familiar rings—diffraction patterns etched into our understanding of quantum mechanics.

"This," Sandeep said, "is the foundation of de Broglie's matter wave hypothesis."

The electron, once a particle, was said to behave like a wave—purely because of what was seen here.

"But what if we never needed to wave it in the first place?"

The structured vacuum wrapped itself around the crystal lattice.

Each atom, now shown to be surrounded by a local field resonance—zones of influence.

The electron approached.

Not as a wave.

But as an energy packet guided by the vacuum's structured pathways.

It didn't spread.

It didn't diffract.

It simply responded to the pre-existing field architecture—releasing energy where conditions allowed, creating interference not from its own shape, but from the shape of space it was moving through.

"Electrons don't become waves," Sandeep clarified.

"The structured vacuum is the wave. The electron is the participant."

The Lattice Pulses – Experiments Rewritten

Each classical experiment now glowed briefly in the amphitheater—side by side:

- The photon, not colliding, but rebalancing.
- The slit, not collapsing, but reorganizing.
- The electron, not diffracting, but responding.

The lattice pulsed—again, again, and then stilled.

Einstein stood. Slowly. Eyes gleaming.

"Structured... not uncertain," he said quietly. "That changes everything."

Bohr didn't speak. He was still looking at the slits—not confused. Just... quieter.

Schrödinger folded his arms. "So the cat was never both. It was just waiting for a reason."

And above them, the old labels faded—"wave," "particle," "collapse"—erased from the air like dust brushed off the edge of an ancient book.

What remained?

Only this:

Structured Energy. Purposeful Interaction. Intelligent Response.

And for the first time in a hundred years—

Light was not confused.

We were.

But not anymore.

Implications for Quantum Mechanics

The amphitheater had grown quiet—but not the kind of silence that comes from absence.

This was cognitive silence. A quiet that follows the collapse of an illusion.

Not of light this time, but of belief.

Bohr sat still. Very still.

The man who once stood on the frontier of thought—who gave form to the idea that nature offers only pieces of truth—was now face-to-face with something he had never fully allowed:

That perhaps nature always told the whole truth... we just kept asking half-formed questions.

His fingers tapped gently on the arm of his seat.

"You're saying," he began cautiously, "that complementarity... was a workaround."

Sandeep nodded. Respectfully.

"A brilliant one," he replied. "But a workaround, nonetheless."

Bohr looked at the structured vacuum hovering in the amphitheater—a field so elegantly deterministic that it made the wave-particle debate feel like a child's riddle.

"So there is no collapse?" Bohr asked quietly.

"Only restructuring," Sandeep replied. "An intelligent lattice adapting to interaction. Nothing collapses. It simply... fulfills purpose."

Bohr leaned back.

The architecture of his entire life's thinking... was still standing.

But it had been remodeled.

And strangely, it felt—relieving.

Not defeated. Just redefined.

Schrödinger exhaled through his nose.

It wasn't a sigh. It was liberation.

The cat had been chasing him for decades. Living and dying in the same breath. Haunting physics and metaphor alike.

Now?

He looked over at the vacuum lattice and smiled.

"The cat was never confused," he muttered. "It was the box that was flawed."

He turned to Sandeep. "You've resolved what I built as a trap. A necessary absurdity to expose a deeper flaw."

"And now?" Sandeep asked.

"Now," Schrödinger said, "I bury the box."

The cat walked away into the vacuum field—tail high, identity intact.

No paradox.

Just poor framing.

And then, Einstein.

He had been quiet for a while. Not from doubt—but from satisfaction.

He stepped toward the center of the amphitheater, toward the lattice, and simply watched.

His voice, when it came, was low. Grounded. Certain.

"I spent half my life refusing the indeterminacy of quantum mechanics—not because I wanted to be difficult—but because I believed that the universe was not built on coin tosses."

He turned to Bohr.

"You asked me to accept uncertainty as truth. I could not."

He turned to Sandeep.

"And now... you've shown us that what we called randomness was simply hidden structure. A vacuum not empty—but organized. Energetic. Intelligent."

He pointed to the beam of light still suspended in the center.

"That is not a mystery anymore."

"That," he said, "is response shaped by context. Not chance. Not duality."

He smiled wider than before.

"And you've done it without rejecting quantum mechanics," he added. "You've simply... completed it."

The New Quantum Paradigm

Sandeep turned to the audience—the thinkers, the listeners, the generations shaped by uncertainty and wavefunctions.

He spoke, not as a challenger of the old, but as the voice of the whole.

"Quantum mechanics gave us the ability to predict. But in doing so, we mistook probability for principle.

We built equations to describe outcomes— but forgot to ask: *what structured those outcomes to begin with?*

The answer was always there—beneath our assumptions.

A vacuum not as a void, but as an intelligent medium. One that balances itself with every action. One that responds not to our questions—but to our intent."

The Amphitheater Evolves

The arena responded—not theatrically, but with grace.

The structured vacuum expanded—calmly weaving through every corner of the amphitheater, forming patterns, then dissolving, like it was rehearsing reality.

Light was no longer flickering with confusion.

It was dancing with clarity.

Bohr stood at last. He approached Sandeep, slower than the others.

He extended his hand.

"I may still defend complementarity," he said, "but I will no longer call it complete."

Schrödinger chuckled. "And I will stop torturing cats."

Einstein leaned in. "It's not about who's right anymore. It's about what *works better.*"

Sandeep smiled—but didn't speak.

Because now, light had already spoken.

And for the first time in a hundred years...

It made sense.

The Lattice of Reality

The amphitheater pulsed again—this time not with tension, but release.

Like the moment a fog lifts.

Like the moment a child finally sees the pattern in a puzzle that once seemed random.

The shimmering structured vacuum that had hovered behind Sandeep began to shift—no longer passive, no longer waiting.

It moved with intention.

But not as a force.

As a *form.*

The light in the arena didn't brighten—it crystallized.

Every ripple of space, every subtle motion of energy, began organizing itself into a breathtaking web of harmony—a grand lattice.

But this wasn't a static grid. It was alive—pulsing with awareness, adapting with grace, breathing through balance.

The amphitheater had become something else entirely.

Not a stage.

Not a battlefield of thoughts.

But a model of reality itself.

And for the first time, the laws weren't written in ink or numbers—but woven into the geometry of space.

Sandeep stepped forward—not as a narrator now, but as a translator.

"This," he said, "is not just a metaphor. It is the mechanism."

He turned slowly, addressing the arena—the thinkers, the watchers, the universe itself.

"What you once called 'particle' and 'wave'—they are not two faces of the same coin.

They are expressions—temporary manifestations—of how structured energy interacts with structured space.

Matter and light are not dual.

They are *directional responses* to purpose.

Each guided, shaped, and stabilized by this invisible architecture—the lattice of equilibrium.

There is no need for collapse. No need for randomness.

Just structured intent."

As he spoke, the lattice responded.

Where once a photon had danced between possibilities, now it flowed in a clear, curved path—no hesitation, no superposition, just choice made visible.

A single electron entered the scene—approaching a crystal lattice. The vacuum field adjusted, reconfiguring in response. The diffraction pattern didn't emerge by accident.

It was engineered.

Like music responding to rhythm.

Like a bird adjusting to wind.

Bohr's double-slit experiment lit up again—but now, the interference was not mystery. It was choreography.

Even the randomness of quantum tunneling now appeared as a calculated redistribution—a field-level negotiation, not a dice roll.

Einstein stared at the unfolding harmony and whispered:

"It was never uncertain. We just couldn't see the blueprint."

Schrödinger traced the pattern of the lattice with his eyes.

"No cat. No box. Just structure adapting to question."

Bohr stepped closer, visibly humbled, yet strangely peaceful.

He extended his hand toward the lattice.

"I've always believed nature reveals only part of herself. But maybe," he said slowly, "she reveals everything—when we finally learn to listen without assumptions."

The lattice pulsed again—soft, slow, complete.

Then it did something unexpected:

It folded inward.

Not as collapse.

But as *resolution.*

Returning to a stillness not of absence—but of *readiness.*

And in that stillness, the greatest paradox of modern science dissolved—not by contradiction, but by context.

Light was not dual. It was directed.

Matter was not confused. It was coordinated.

And the universe was not uncertain.

It was structured.

Chapter Closure

The lattice began to dissolve—not in loss, but in fulfillment.

Its glow faded gently, like a song reaching its final note, not vanishing, but settling into memory.

Piece by piece, the crystalline geometry of the structured vacuum folded back into the invisible—the amphitheater restoring itself brick by brick, echo by echo.

But it was not the same.

The floor was still marble. The sky, still limitless. The pillars stood tall.

Yet everything had changed.

Because now the arena did not stand atop theories. It stood atop understanding.

Bohr rose first.

He nodded—not just to Sandeep, but to the field itself, as though finally seeing a friend he never knew was there.

He walked away—not defeated, but transformed. The need to defend duality had melted into the satisfaction of wholeness.

Schrödinger followed, lighter now, free of the cat, the box, the paradox. His thoughts were no longer trapped in probabilities. They flowed with structure, danced with meaning.

Einstein lingered, looking once more at the place where the lattice had hovered. Then he smiled—peaceful, rare, complete.

He had spent his life searching for order beneath the chaos. Now, he had seen it.

And this time... he didn't need to write it on a blackboard. He carried it within.

Sandeep remained alone in the center.

Not alone, truly—but accompanied by silence, the kind that only appears when something *real* has been revealed.

He looked around, slowly, absorbing what had just unfolded.

"Light was never uncertain," he whispered. "It was never torn between two forms." "It was always whole—structured, waiting for us to ask better questions."

He raised his eyes toward the horizon of the amphitheater, where the shimmer of structured vacuum still flickered faintly—like a promise.

"By embracing Structured Vacuum Energy," he said, "we transcend dualities.

We stop slicing nature into labels—particle, wave, force, collapse.

And instead, we begin to witness the coherent fabric of reality—not fractured by observation, but revealed by purpose."

The wind shifted—soft and knowing.

The amphitheater glowed once more, resetting itself, preparing.

Because the next transformation was near.

And this time, the stage wouldn't be set by paradox—but by pattern.

Not by contradiction, but by spectrum.

A new resonance began to rise—not a question, but a field of possibilities.

From the softest whisper of intention to the densest form of matter, energy would now reveal itself—not in isolation, but in its entire range.

Chapter 6: The Spectrum of Reality

Forms of Energy and Their Role

Hot Seat: Nikola Tesla, James Clerk Maxwell

Wild Card Entry: Srinivasa Ramanujan (The Poetic Mystic)

The light had spoken.

And now, the silence after its voice vibrated through the amphitheater—not as emptiness, but as resonance.

A new hum stirred within the stones, a soft tremor that wasn't heard with ears, but felt through presence. It wasn't time, or thought, or space—It was frequency.

The walls of the arena dissolved again, but this time, they didn't disappear into chaos or cosmic abstraction.

They shifted into gradients—wavelengths, amplitudes, invisible spectra made visible. A sea of colors pulsed through the air—infrared, ultraviolet, magnetic, sonic—each not a different force, but a different face of the same presence.

The audience didn't just observe anymore. They vibrated with the room.

What was once a stage had become a spectrum. And standing in the middle of it all—emerging not with footsteps, but with a rising frequency—was Sandeep.

He held no artifact.

He carried no formula.

He stood still—because the truth no longer needed a container.

It only needed a range.

And then—Tesla arrived.

Like lightning shaped into a man.

His presence crackled—quietly. Powerfully.

He didn't greet. He didn't ask.

He simply said:

"If you wish to understand everything... understand *vibration.*"

Moments later, space folded in elegant arcs of invisible lines, and Maxwell emerged—not as a man, but as a field equation made flesh.

He moved with symmetry, spoke with certainty, and radiated balance.

"Electricity and magnetism," he said, "are not separate. They are in phase. A single pulse—divided only by your understanding."

And just when the amphitheater began to absorb their presence, a breeze of numbers swept in.

Soft. Warm. Sacred.

A barefoot man walked silently between dimensions.

Ramanujan.

No announcement. No applause.

Only symbols glowing in his wake.

He didn't speak. He hummed a number—And the air around him began to balance.

The wild card had entered.

But he did not come to debate.

He came to reveal the language of energy that even equations dare not speak.

And thus, the spectrum had gathered.

Not to compete.

But to resonate.

Because this chapter was not about what energy *is.*

It was about how it expresses.

And why every expression—seen or unseen—still sings the same song.

The Arena of Vibration

Reality didn't crack or collapse this time.

It sang.

The transformation of the amphitheater began without warning, without drama—only rhythm.

Walls dissolved, not into void, but into vibration—soft pulses that rippled outward, not just through space, but through understanding.

The air thickened—not with substance, but with pattern.

No longer bound by marble and matter, the amphitheater became a living field—shifting gradients of frequency stretched across every axis.

Light glowed not from lamps, but from intentions. Sound danced, but no one spoke. Color emerged where thought aligned with movement—vibrant blues, deep reds, radiant golds, flowing like magnetic rivers through space.

Above, invisible field lines arched and crossed—visible now to the intuitive eye. Electricity wasn't a current—it was a ripple in the silent song. Magnetism no longer pulled—it embraced.

The audience didn't observe this world.

They *resonated with it.*

Their thoughts, their presence, even their breath began to sync with the oscillations around them.

Every heartbeat was part of the hum. Every motion tuned to a field unseen, yet utterly felt.

The arena had become a harmonic organism.

There was no chaos here.

No conflict.

Only the geometry of coherence.

And then—he arrived.

Not as a figure. But as a frequency.

From the center of this orchestral lattice, Sandeep emerged—not by walking forward, but by condensing out of resonance.

A subtle swirl of golden threads converged, slowed, formed edges, volume, weight—until a human form stood where moments before, there was only wave.

He didn't say a word.

His presence alone told the story:

That form is not fundamental. It is resolved vibration. Matter is not separate from field. It is the *pause* in a song... that allows the note to exist.

The audience didn't applaud.

They *stabilized*—their frequencies aligning with his presence.

And in that harmony, the chapter began.

A journey into the spectrum of energy—

Not as a scientific abstraction, But as the true language of reality.

Because what came next was not about discovering forms...

But about feeling them.

Introduction of Thinkers

The arena hummed like a tuning fork touched by thought.

It no longer waited for words.

It awaited resonance.

And then—like lightning remembering its path—he appeared.

Nikola Tesla

There was no sound when Tesla entered.

Only a spark.

A filament of thought, a coil of subtle charge, winding through space until it coalesced into form.

Tall. Composed. Draped in a tailored black suit that shimmered faintly with static.

His presence wasn't loud—it was high-voltage elegance.

He didn't move. He *oscillated.*

Electricity danced at his fingertips—not for show, but because it couldn't help it. He was its language.

He surveyed the vibrational field around him—not surprised, but satisfied. As if he had been here before.

"If you want to understand the universe," Tesla said, his voice barely louder than the resonance around him, "think in terms of frequency and vibration."

"Not particles. Not equations. Not probabilities."

"Just... pulse."

He raised a hand and the entire arena responded—colors aligning, tones harmonizing. He wasn't just in tune with the space.

He was tuning it.

James Clerk Maxwell

From the curved lattice above, field lines began to bend inward.

Not in chaos, but in perfect proportion.

And from those lines, a man emerged—grace wrapped in mathematics.

James Clerk Maxwell stepped forward like a geometric proof turning into flesh.

His posture was exact. His movements clean. His aura glowing with the rhythm of equations made visual.

He moved through the amphitheater as if every step had already been calculated and perfectly allowed by the field.

Behind him, curved lines of electric and magnetic flow shimmered—dancing with orthogonal grace.

He did not shout his truth. He demonstrated it.

"What you call electricity," he said, pointing to a luminous line, "and what you call magnetism," pointing to another, "are not opposites. They are in rhythm. One chasing the other."

"Together, they form light—not just as waves, but as an intention in motion."

His presence wasn't dramatic.

It was discipline given form.

The architect of electromagnetic harmony had arrived—not to argue, but to complete the chord.

Wild Card Entry: Srinivasa Ramanujan

There was no light. No sound. No motion.

Only a number.

It hovered in the air: 1729.

And then... a barefoot step.

Soft. Silent. Timeless.

Ramanujan emerged like a poem finishing itself.

Not through thought.

Through *feeling.*

He walked gently across the lattice—not disturbing it, but syncing with it. Every number he whispered shimmered in the air, not as digits, but as forms.

∞. ϕ. π. e. They floated. They pulsed. They turned.

He did not greet the others.

He greeted the field.

"The cosmos," he murmured, "is a function of beauty, not utility."

"Every equation is a mantra...and every solution, a sacred breath of balance."

He looked toward Sandeep—not for permission, but in recognition.

And with his fingers, he drew spirals into the air.

Golden spirals.

Harmonic curves.

Paths of energy... that never ended.

Ramanujan would not argue.

He would not debate.

He had arrived not to explain.

But to reveal what equations alone could not say:

That energy is not just physics. It is poetry—recited by the universe in its own dialect of silence.

Three thinkers.

Three frequencies.

One spectrum.

And for the first time, science, structure, and mysticism stood not as rivals, but as notes in the same scale.

The arena pulsed in agreement.

Because now—at last—

It was time to classify everything.

Not by what it was...

But by how it resonated.

Classification of Energy in UED

The Three States of Energy: Stable – In Action – Condensed

The amphitheater had stabilized—not into stillness, but into readiness.

Three great minds stood like tuning forks, each vibrating at their resonant truth. And at the center stood Sandeep, preparing to unify it all.

He turned to the audience—not to instruct, but to awaken.

"Energy," he said, "is not one thing. It is one intelligence, seen in three distinct states—each a form of balance or its disturbance. Not forces. Not particles. Forms. of. Being."

He raised his hand, and the arena divided itself—gracefully—into three spectral zones:

A glowing stillness to the left, a vibrating field in the center, and a dense, humming solidity to the right.

1. Stable Energy – The Silent Source

The left zone shimmered—but did not move.

Not a wave. Not a flash. Just presence.

This was not emptiness.

This was perfect equilibrium.

The structured vacuum—unshaken, unprovoked.

A silent sea, capable of everything but needing nothing.

Sandeep gestured to it.

"This is Stable Energy. Not seen. Not heard. But infinite in potential.

It does not act. It waits."

The audience stared, expecting change—but none came.

That was the point.

Then, softly, like a breeze through a monastery, Ramanujan stepped forward.

His eyes half-closed, his breath measured.

He pointed to the shimmering stillness and whispered:

"Like zero...Full. But unseen."

And just for a moment, the number 0 hovered above the field—not hollow, but glowing. Alive.

2. Energy in Action – Vibration, the Middle State

The central zone pulsed—wave after wave, echo upon echo.

A field of pure interaction.

This was energy that had heard a call. It was no longer still. But it wasn't matter either.

It was in motion—not chaotic, but calculated.

"This," Sandeep said, "is Energy in Action. The moment equilibrium is disturbed, the field begins to oscillate."

"Waves form—not because energy *wants* to wave, but because the field is seeking balance."

Tesla stepped forward now, eyes gleaming, coat sparking at the cuffs.

He didn't look at the audience.

He looked *through* them.

And he said:

"Vibration is the soul's signature.

To exist is to vibrate.

To live is to tune yourself into the field."

He raised a hand and the waves aligned—forming a spiral of color, frequency, and intent.

Resonance became visible.

Purpose became pattern.

3. Condensed Energy – Matter as Resistance

To the right, the third zone thickened.

Not in color, but in density.

It didn't ripple. It didn't hum.

It held.

This was energy not moving freely, but trapped within its own tension.

Structured. Defined. Mass.

"And this," said Sandeep, "is Condensed Energy. Where resistance becomes identity. Where imbalance reaches such intensity... that it must hold itself together."

Matter wasn't a thing.

It was a stubborn wave.

A loop so tight, it couldn't escape itself.

Maxwell walked into the field with poise—his coat flowing like field lines in motion.

He touched the density—gently—and turned to speak:

"Matter is not substance. It is structured tension—a rhythm that has folded into stability."

He drew invisible lines—electromagnetic, geometric, graceful—and the field responded.

Every atom in the amphitheater pulsed with recognition.

They were not solid.

They were held.

The Spectrum Realized

The arena now displayed it all—side by side.

- Stable Energy – the unshaken potential.
- Energy in Action – the seeking wave.
- Condensed Energy – the held form.

Each state not separate—but transitional.

"All forms of existence," Sandeep said, now speaking with soft power, "are energy in different moods.

What you call an object...

What you call light...

What you call self...

Are simply positions on this spectrum."

And for the first time, the amphitheater was not observing a theory.

It was feeling the universe organize itself.

It was not about mass, or force, or frequency anymore.

It was about relationship.

And the dance between balance and intention.

Redefining Matter, Waves, and Potential

The amphitheater now resonated with understanding.

But Sandeep wasn't finished.

The structure had been revealed.

Now, the misinterpretations needed to be corrected—gently, precisely, like tuning an instrument that had long been played out of key.

He walked between the three zones of energy, his presence shifting as he moved—from stillness to vibration to weight.

Matter – The Illusion of Solidity

He stood in the field of condensed energy—its density almost audible.

He extended his hand, and a solid object formed—a sphere, seemingly made of polished stone.

The audience instinctively recognized it: *matter.*

"We've been taught," Sandeep began, "that matter is 'real'—the ultimate building block of reality."

"But matter is not a starting point. It is a result."

He rotated the sphere midair.

It began to shimmer, revealing its layers.

Beneath the surface was motion.

Tightly bound, looping, resisting—frozen rhythm.

"Matter," he continued, "is not a separate category of existence.

It is energy... compressed by structure into holding itself."

"It is not solid. It is stubborn."

Maxwell nodded subtly, tracing field lines that wrapped around the sphere, revealing the electromagnetic scaffolding that gave it form.

Matter, at last, was seen not as substance—but structured equilibrium under pressure.

Waves – Misunderstood Messengers

Sandeep moved to the center now—where the field still pulsed with motion.

A soft wave formed in the lattice—gliding, clean, coherent.

"We say that energy *is* a wave. But that's not true."

"A wave is not energy itself.

A wave is what happens when energy interacts with structure."

"It's a *behavior*, not a thing."

The wave was traced from origin—rising from a single point of imbalance in the vacuum field.

It rippled not randomly, but intelligently—adjusting direction, splitting at junctions, curving around resistance.

"Waves," Sandeep explained, "are energy in conversation with the field.

They seek balance, not escape.

And when they fail to restore it, they condense... into matter."

Tesla raised a hand and resonated with the wave. It shifted slightly in rhythm—acknowledging him.

"Frequency," Tesla said softly, "is not decoration.

It is the voice of the field."

The amphitheater pulsed in agreement.

Potential Energy – A Misnamed Stillness

Then Sandeep paused—and stepped into the silent zone again.

He pointed to a still point in the field—no motion, no wave, no visible energy.

Yet, as he moved his hand closer, the point began to stir—gently, like a thought being remembered.

"We call this potential energy—as if something is 'stored' inside an object. But nothing is stored."

"There is no hidden pile of energy waiting inside a spring or a rock."

"There is only pre-configured balance."

A visual unfolded:

- A wave formed in the lattice.
- It curled inward—tightening, slowing, until it became a dense knot.
- The knot held, perfectly stable—not moving, but ready.
- Then, a slight disturbance—and it unfolded again into motion.

The loop was complete:

Wave → Particle → Wave

Not a transformation.

A cycle.

"Potential energy," Sandeep said, "is not about what is stored. It's about what is possible—given the current structure."

Ramanujan whispered, as if decoding the idea with one sacred number:

"Potential is not fuel.

It is readiness in balance.

Like π—it never ends... but it begins with stillness."

The Visual Symphony – Energy's True Motion

Suddenly, the amphitheater dimmed.

A holographic performance began:

- A point of energy appeared in the vacuum.
- It vibrated, creating a wave.
- The wave curled, tightened, slowed.
- A particle formed.
- The field restructured itself to contain it.
- Then, with slight disturbance—the particle unraveled back into motion.

This was not magic.

This was the unified motion of the cosmos.

A breathing cycle:

From silence → to motion → to mass → back to motion → toward silence again.

Not three different things.

Just one truth in three expressions.

The audience sat still—not because they didn't understand, but because they finally did.

And as Sandeep stepped away, the three thinkers remained in place—Tesla humming with resonance, Maxwell sketching invisible diagrams, Ramanujan weaving equations into golden spirals.

Together, they had shown the world what energy really was:

Not stored. Not split. Not confused.

But always...in flow.

Ramanujan's Symbolic Vision

The amphitheater had become an orchestra of resonance.

Tesla's frequency had tuned the vibration. Maxwell's fields had mapped the invisible. Sandeep had classified the flow.

But now... something deeper stirred. Not a theory. Not a structure. A feeling. A song.

And from the soft center of silence, Ramanujan stepped forward.

He held no chalk. No device. Just intention.

His eyes glowed—not with knowledge, but with knowing.

He didn't explain. He intoned.

He raised a hand.

A number hovered in the air: 0

It pulsed, not with absence, but completeness.

"This is Stable Energy," Ramanujan whispered. "Nothing. Yet total.

Not silence—potential uninvoked.

Zero is not empty. It is unmoved."

The audience felt it—not as concept, but as vibration without direction. The vacuum before the first breath.

Then... he moved.

A second number rose: 1

This time, it didn't glow—it vibrated.

A single line, a lone pulse, the first ripple in perfect stillness.

"One is the first imbalance," he said softly. "It is the *choice to be seen.*

It breaks silence.

It begins rhythm."

Tesla nodded silently. He recognized this as the origin of vibration.

Then Ramanujan continued.

With a sweep of his hand, a symbol flooded the air: ∞

The lemniscate glowed blue—pulsing from both ends. A loop that never closed. A motion that never rested.

"Infinity," Ramanujan chanted, "is unresolved imbalance.

Motion that forgets to return. Energy that flees from rest."

Maxwell lowered his eyes.

This was not disorder—it was excess. The runaway wave. The field unanchored.

Then, in silence, Ramanujan drew his final symbol: Φ

The golden ratio. But it didn't stay flat. It breathed.

The golden spiral emerged—growing outward, then folding back in, tracing balance within imbalance, motion within stillness.

"This," Ramanujan said, his voice almost a song, "is cosmic equilibrium.

The divine proportion where form finds rest within movement."

The spiral wrapped through the lattice, nested within itself, fractaling across dimensions.

Tesla closed his eyes.

Maxwell stepped back.

Because neither math nor energy could compete with this.

Equations Beyond Language

Ramanujan extended both hands now.

Equations filled the air—not with operators, but with *intention.*

They were not to be solved.

They were to be felt.

- Equations spun like mandalas.

- Numbers glowed as mantras.
- Prime sequences bent into golden arcs.
- Constants began to hum.

These were not formulas.

They were frequencies encoded in form.

He did not read them aloud.

Because he didn't need to.

Each observer saw what they were ready to see.

Each thinker felt the one equation that belonged to them.

Sandeep saw the equation for structured balance.

Tesla saw the ratio of perfect resonance.

Maxwell saw the symmetry that held opposites in phase.

Ramanujan saw nothing. Because he had become the numbers.

And then, the poetic silence.

He stepped back, letting the lattice absorb his equations.

Not erasing them.

Embedding them.

"The lattice does not forget," he whispered. "Every structure is a remainder of intention.

Every form... is memory."

The equations faded, but their vibration stayed—woven into the amphitheater.

Tesla and Maxwell said nothing.

Because at that moment, nothing more could be said.

There was no debate.

No doubt.

Just the echo of a truth too sacred for language.

Ramanujan smiled—almost like a child who had just released a secret to the sky.

He didn't bow. He didn't stay.

He dissolved into the lattice, leaving behind only a golden curve... still spiraling...still balanced...still singing.

The Color of Energy

The amphitheater began to hum—not with sound, but with light that could be heard.

Frequency thickened. The air shimmered.

And then, like the universe turning a page, the entire arena flooded with color.

Not painted. Not projected. Revealed.

From every curve of the structured vacuum, bands of visible frequency unfolded—waves no longer hidden, but radiant.

- Infrared moved slow and warm, like breath in deep meditation.
- Ultraviolet zipped like thought leaping ahead of language.
- And in between—the full spectrum—flowing, converging,

expanding.

It was no longer science.

It was symphony.

Each hue told a story:

Red glowed low, close to the earth. Dense. Heavy. Rooted.

"This," Sandeep explained, "is condensation.

Where energy slows... coils in on itself... becomes form."

Atoms. Bodies. Planets.

All born in red's patient curve.

Blue shimmered fast, sharp, and high—like clarity itself.

"Here," he continued, "is acceleration.

Energy in action, seeking alignment, resonating into motion."

This was light in flight, thought in momentum, intention in progress.

White hovered in the center—soft, glowing, complete.

Not a color, but all colors.

"And this," Sandeep said gently, "is resonance in balance.

Stable energy. The origin, the return, the knowing."

Ramanujan's golden spiral reappeared inside it—reminding everyone: even color has memory.

And then, a transformation.

The spectrum began to bend—not in rainbow arcs, but in conscious alignment.

- Red condensed into spheres.
- Blue stretched into waves.
- White formed lattices—anchoring the rest.

It was no longer the electromagnetic spectrum.

It was the existential spectrum.

A visual declaration that every aspect of reality—form, force, emotion, thought—is simply energy occupying a different place in the field.

Sandeep stood at the center of the swirling frequency—a silhouette between red and blue, bathed in white.

His voice was not raised.

It *resonated.*

"You are not made of matter," he said. "You are not built from particles or pulled by forces."

"You... are a position—on the spectrum of structured energy."

"You are a note...

in the universe's infinite scale."

The amphitheater didn't erupt.

It aligned.

Because alignment, not applause, is the true sign of understanding.

Every frequency now pulsed in harmony—not uniform, but coherent.

Every being present found their place—not by hierarchy, but by resonance.

And for a brief, brilliant moment—Existence itself was in tune.

Chapter Closure

The colors faded—not into absence, but into balance.

No longer swirling. No longer shouting.

Each frequency found its resting place in the invisible lattice, humming softly in the background—like a sacred instrument set down after the final note.

The amphitheater had returned to its most natural state:

Not empty. Not still.

But vibrating at perfect equilibrium.

The field held no tension.

No question.

Only awareness.

A silence filled with knowing.

Tesla was the first to leave.

He didn't walk—he flowed.

A ripple of light passed through his frame as he began to dissolve into his own resonance, like a current reconnecting to its source.

He smiled—not at Sandeep, not at the audience, but at the field itself.

"Now," he said, his voice a whisper of voltage, "let them feel it."

And just like that—he was gone.

But the charge he left behind still shimmered in the amphitheater air.

Maxwell remained a moment longer.

He stood before his own field equations, suspended midair—beautiful, symmetrical, precise.

But now he saw them differently.

Not as answers.

As approximations of intention.

He gently reached out, and with a stroke of his hand, he erased them.

Not with regret—but with reverence.

He turned to Ramanujan, nodded, and bowed.

A scientist surrendering to the sacred.

A master of fields honoring the mystic behind the veil.

Ramanujan never said goodbye.

He was mid-sentence, tracing a spiral in the air when his voice simply faded—like a mantra returning to silence.

His body dissolved—not into light, not into dust—but into pattern.

And there, where he stood, a string of numbers floated in the air.

No explanation. No label.

Just a glowing sequence, pulsing in golden rhythm, as if still singing in the language of the cosmos.

Those who understood... smiled.

Those who didn't... still felt peace.

And Sandeep?

He remained. Not speaking. Not explaining. Just quietly resonating.

His heartbeat now synchronized with the lattice.

His thoughts woven into the field.

Because he knew:

This was not the end of the journey.

This was only the settling of the surface.

Beneath the spectrum... lay the source.

Beneath form and frequency... the first impulse.

"Now," he whispered to himself, "we go deeper."

"From the spectrum of reality...to the seed of individuality."

The arena grew dim.

Not with darkness. But with introspection.

The structured vacuum was still watching.

And the next chapter would begin not with a color...But with a choice.

Epilogue

The Arena Breathes

The amphitheater stood—not in silence, but in absorption.

It no longer echoed the clang of intellectual swords. No longer sparked with the friction of opposing thought.

Now, it simply held.

Held the weight of questions no longer unanswered—just understood differently.

What began as a chamber of challenge...had become a mirror.

A mirror reflecting not reality itself, but the fractures in how we once tried to know it.

Truth had been unshackled—from absolutes and authority. It had shed its robes of permanence and stepped forth as a living outcome, shaped by intent, not doctrine.

Gravity, once hailed as the grand puppeteer, had bowed in humility—revealing itself not as a force, but as a symptom of imbalance. An effect... not a cause.

Time had melted—its clocks shattered not by chaos, but by clarity. What remained was not a ticking arrow, but a field of causal experience—measured by memory, not mechanics.

And the origin—the fabled beginning, the singularity, the first breath of everything—had dissolved into structure.

Not a point in space.

But a principle in pattern.

The thinkers who once filled this stage—Plato, Einstein, Bohr, Hawking, Penrose, Tesla, Maxwell, Schrödinger, Ramanujan—were now still.

Their battles done.

Their boundaries redrawn.

Not one of them had been disproven.

But each had been expanded—lifted from the realm of "what is" to "why it appears that way."

Their voices no longer echoed across the chamber. They echoed within it.

Softer now.

Wiser.

At the center, Sandeep stood alone.

Not as a victor. Not as a prophet. But as the last listener in a room where the universe had spoken.

He didn't need to raise a hand or speak a word.

The amphitheater breathed with him.

Together, they exhaled a century of confusion. Together, they inhaled a future of coherence.

Above him, the structured vacuum shimmered once more.

Not in mystery. But in readiness.

It had revealed itself—not as emptiness, but as the sacred stage upon which energy, matter, light, and thought all played their parts.

And now, it waited. Because in that shimmering stillness, something shifted—

Not in direction. But in dimension.

Not in space. But in subject.

The cosmos had been questioned.

Its illusions cracked. Its laws reframed.

But the most volatile, unpredictable, misunderstood force in the known universe... still stood untouched:

The human being.

What force bends our will? What field shapes our choices? What disturbance condenses into consciousness?

The amphitheater had hosted the great debates of physics, space, time, and energy.

Now, it was preparing for something far more intimate.

More chaotic. More divine.

The mind. The intellect. The architecture of choice.

And so, as the stage emptied of equations and spirals, as the last color of resonance dissolved into silence...

The arena did not close. It turned inward.

And Sandeep followed it.

Toward the next crack in the known—The Human Game.

Afterword

Cracks Revealed, Truth Reimagined

By Sandeep

This book was not written to disprove. It was written to reframe.

Not to discard science—But to realign it with something older than equations, and far more stable than consensus: The deeper intelligence of coherence.

For too long, we've worshipped formulas while ignoring the fabric behind them. We've memorized definitions but forgotten how to define for ourselves. We've built entire realities around gravity, time, force, matter—Yet few dared to ask:

What if we misunderstood them all?

Not because science failed—But because it stopped asking why.

Cracks in the Known is a journey through those foundational illusions:

- Where truth is no longer a fixed monument, but a living outcome of perspective and context.
- Where gravity is not an invisible hand pulling us down, but a balancing act of structured energy fields in tension.
- Where time doesn't tick forward but loops through causal awareness.
- Where light isn't dual—it's decisive, responding to our intent, not just our instruments.
- And where the origin of everything... may not be a point in space or time, but a principle encoded in equilibrium itself.

I didn't write this to confuse the reader.

I wrote it to un-confuse them.

Because I believe many of us—students, scientists, philosophers, even casual thinkers—have always *felt* the cracks. But we were trained to ignore them. To accept contradiction as mystery. To worship math as meaning.

But mystery is not meaning.

Mystery is invitation.

This book is my answer to that invitation.

A philosophical rebellion against mental laziness. A respectful confrontation with giants. A silent handshake with truth that lives beyond textbooks.

Each chapter is not a conclusion—It's a doorway.

And beyond these doors lies not uncertainty...But clarity reframed.

These cracks matter—not just for physics, but for philosophy. Not just for equations, but for ethics. Not just for discovering the universe—But for reclaiming the self.

Because if light responds to intent, If time bends to memory, If reality organizes itself based on structured energy and choice...

Then what does that make *you*?

This is only Part I.

We've dismantled the external.

Now, we descend inward—Into mind, will, identity, and illusion.

From *Cracks in the Known* to *The Human Game*, The war of thoughts is just beginning.

Stay with me.

Let's keep cracking.

Timeline of Thought

A Journey Through the Arena

What follows is not a table of contents.

It is a map of transformation.

Each chapter in *Cracks in the Known* was not just a lesson—it was a confrontation. A conversation. A crack in certainty.

Here is the path we walked together—one debate at a time.

Chapter 1: The Idea of Truth

Hot Seat: *Plato*

The arena awakens. Truth is no longer absolute—it is an evolving consequence of perspective, context, and purpose.

Plato defends the ideal, but the ideal begins to breathe.

Chapter 2: Gravity – Force or Fiction?

Hot Seat: *Isaac Newton & Albert Einstein*

The pillars shake. Gravity, once the most trusted law, is reframed as a consequence—not a force, but a field response to imbalance. Newton speaks of attraction. Einstein of curvature. But both step aside as the arena reveals balance, not pull.

Chapter 3: The Time Trap

Hot Seat: *Stephen Hawking*

The clock melts. Time is no longer a flowing river—it is a reflection of causality, memory, and motion within structure. Hawking argues for arrows and entropy, but the arena whispers: *What you measure as time... is simply change you can't ignore.*

Chapter 4: The Origin Debate – Was There a Beginning?

Hot Seat: *Georges Lemaître & Roger Penrose*

The beginning dissolves. Big Bang becomes breath—cyclic, intelligent, structured. The question of origin is not about when it began, but *why we think beginnings are necessary.* Penrose proposes cycles. Lemaître defends creation. But Sandeep introduces *Structure as Continuity.*

Chapter 5: The Duality Dilemma – What is Light, Really?

Hot Seat: *Niels Bohr, Erwin Schrödinger, Albert Einstein*

The photon stands trial. Is light wave or particle? The thinkers clash—Bohr with complementarity, Schrödinger with paradox, Einstein with skepticism. But then comes the Structured Vacuum—revealing light not as a thing, but as a response to purpose.

Chapter 6: The Spectrum of Reality – Forms of Energy and Their Role

Hot Seat: *Nikola Tesla, James Clerk Maxwell* Wild Card Entry: *Srinivasa Ramanujan*

The field sings. Energy is not singular—it exists in states: stable, active, condensed. Waves are conversations. Matter is held vibration. Potential is balance waiting for a reason. Ramanujan speaks in symbols, Tesla in frequencies, Maxwell in curves. Together, they unveil the spectrum—not just of physics, but of existence.

Each chapter was a layer peeled. Each thinker, a reflection of our own clinging to certainty. And with each debate, the arena cracked—not into chaos, but into coherence.

What remains is not less... but more refined.

Truth has bent, but not broken.

Science has shifted, but not shattered.

And what lies ahead... is the war within.

Notes

Universal Energy Dynamics (UED)

A Primer for the Curious, the Thinkers, and the Unsettled

If this book felt like it was dismantling reality and reassembling it into something more coherent—you're not imagining it.

What you've been experiencing through every chapter is not just a new perspective on science—It's a new framework.

That framework is called **Universal Energy Dynamics (UED)**.

What is UED?

Universal Energy Dynamics (UED) is a unified philosophical-scientific lens that redefines the building blocks of the universe—not as disconnected forces or particles, but as structured responses within a living field of intelligence.

In UED:

- Energy is not a property—it is a presence.
- Matter is not a substance—it is condensed structure.
- Waves are not independent—they are the negotiation between imbalance and balance.
- Observation is not passive—it is participatory.
- And most importantly—
- Nothing is random. Everything is relational.

The Core Principles of UED

1. The Structured Vacuum is Real

The "vacuum" of space is not empty. It is intelligently structured—a field capable of organizing, balancing, and responding.

This is known as Structured Vacuum Energy (SVE)—an ever-present substrate from which all forms arise.

2. Energy is Structured Intent

Energy is not just a measurable quantity. It is a manifestation of purpose within structure.

What we see as force, light, or motion...Is energy fulfilling a role in a field attempting to restore balance.

Energy doesn't "flow"—It responds.

3. Spectrum Determines Identity

Everything that exists—light, sound, thought, emotion, matter—is simply energy occupying a position on a structured spectrum.

That position defines what it appears to be:

- Stable Energy → Unseen potential
- Energy in Action → Waves, fields, interactions
- Condensed Energy → Matter, mass, particles

You are not made of matter.

You are a precise location on the spectrum of structured energy, defined by intention and response.

Want to Explore More?

If UED intrigued you—or if you're still wrapping your mind around what just cracked open—you may enjoy diving into these companion works:

Vishwarup Darshan – The Complete Vision of Reality

A philosophical blueprint of the universe using the lens of UED, exploring how everything—seen and unseen—emerges from structured energy and dissolves back into it. It connects science, cosmology, and sacred geometry.

Energy in Action – The Soul Demystified

A deep look at how energy moves within life itself. What we call soul, spirit, thought, and awareness—are decoded through UED, linking biology with cosmology, and psychology with physics.

These works build the philosophical and spiritual foundation on which *Cracks in the Known* stands.

Where UED is Heading

UED is not a theory to be memorized. It is a lens to live through.

Future books in *The War of Thoughts* series will continue to apply and expand UED into:

- Psychology and Will (*The Human Game*)
- Ethics and Karma (*The Karma Paradox*)

- Creativity and Intelligence (*The Untold Story of Intelligence*)
- Origin, Self, and the Divine Field (*Manifestation*)

This isn't just about science.

It's about reclaiming the mind that uses science.

And UED is the compass.

What's Next

Part 2 – The Human Game

You've questioned the universe. Now it's time to question the one who's been watching it.

Cracks in the Known pulled apart the illusions of time, gravity, light, and origin. It challenged the foundations of science—not to destroy them, but to return them to truth.

But even as the cosmos surrendered its secrets, one question remained untouched: Who is asking all these questions?

That is the puzzle of Part 2.

From Cosmos to Consciousness

In *The Human Game*, the arena shifts—no longer outward, but inward.

No galaxies. No black holes. No fundamental forces.

Instead, we confront something far more unpredictable:

- Will.
- Intellect.
- Memory.
- Morality.
- Choice.

This is not the universe outside. This is the universe within—and it is more chaotic, more layered, and far more dangerous.

New Thinkers. New Terrain.

The amphitheater returns—transformed into a psychological mirror.

And onto its stage step new minds—giants of introspection and inner complexity:

- Sigmund Freud – Unveiling the unconscious as a battlefield of instincts.
- Carl Jung – Speaking in symbols, myths, archetypes, and collective dreams.
- Friedrich Nietzsche – Ripping apart moral comfort with unapologetic existential fire.
- Krishna – The cosmic counselor, questioning duty, karma, and dharma in the middle of moral war.

And Sandeep?

He stands not as physicist, but as witness—a challenger of the human condition, holding the mirror steady when it begins to crack.

The New Battleground

The war of thoughts now enters your head—and your history.

Here, the battlefield is not spacetime.

It's the architecture of the self.

- How do we decide what's right?
- Is morality absolute—or self-constructed?
- What governs will? And can intellect truly be free?
- Are we shaped by memory, or do we shape it?
- Is consciousness a signal... or a story we tell ourselves?

There are no lab instruments here. Only emotion. Instinct. Belief. Guilt. Hope. And karma.

A Deeper Tone, A Sharper Edge

If *Cracks in the Known* was philosophical and expansive, *The Human Game* is personal and surgical.

It will not debate your theories.

It will question your intentions.

It will not ask what the universe is made of. It will ask why you act the way you do—even when you know better.

This is where science meets self-deception. Where knowledge meets ego. Where free will stares into its own reflection—and doesn't like what it sees.

The War Continues

"We've cracked the universe." "Now, we must crack ourselves."

The Human Game begins soon.

The players are ready. The rules... are already broken.

Acknowledgments

This book was never meant to be written. It emerged—not as a project, but as a necessity. A quiet revolt against assumptions, a response to the itch that reason alone could not scratch.

To the thinkers whose voices I challenged in these pages—Plato, Einstein, Hawking, Bohr, Lemaître, Penrose, and many more—you were never opponents. You were companions in a journey that stretched across time, disciplines, and dimensions. Thank you for daring to think in the first place.

To the universe—thank you for staying silent long enough for us to misinterpret you. In that silence, we found our questions. In those questions, we discovered ourselves.

To my family, who watched me drift into thought and return hours later with fragments of another world—thank you for your patience, love, and grounding presence.

To those who have read this book not to agree or disagree, but to **think**—this was always for you. May your thoughts never settle, and may your silence always speak louder than time.

And finally, to the stillness behind all beginnings—the Structured Vacuum from which all things arise—this book is my ripple back to you.

— *Sandeep*

About the Author

Sandeep Chavan is an independent thinker, educator, and writer who challenges conventional wisdom through a unique blend of science, philosophy, and consciousness studies. With a background in engineering and physics, he has spent decades dismantling dogma—whether scientific, spiritual, or societal—and reframing foundational concepts through a lens of clarity, balance, and insight.

He is the creator of **Universal Energy Dynamics (UED)**, a framework that redefines force, time, gravity, and existence itself—not as isolated phenomena, but as consequences of energy seeking equilibrium through intelligence. His work spans cosmology, metaphysics, psychology, and the nature of human awareness.

Sandeep writes not to convince, but to awaken. His books are not explanations; they are invitations—to pause, to question, and to remember that the universe may not be what we've been told... and neither are we.

Cracks in the Known is the first volume in The War of Thoughts series—a global, timeless debate between humanity's greatest minds, revisited with bold inquiry and fresh eyes. The battle is not for victory, but for understanding.

When he is not writing, Sandeep continues to teach, explore, and quietly dismantle reality—one assumption at a time.

Disclaimer

This is a work of philosophical imagination and intellectual exploration.

While this book draws upon historical figures, scientific theories, and philosophical frameworks, the dialogues and interpretations presented herein are entirely fictional and represent the author's creative engagement with foundational questions. The legendary thinkers portrayed—such as Einstein, Plato, Buddha, Newton, and others—are used symbolically, not biographically, to explore concepts that challenge conventional understanding.

The scientific, metaphysical, and psychological arguments put forth in this book are not intended as established academic claims or empirical declarations. They are thought experiments—offered to provoke reflection, curiosity, and independent inquiry. Readers are encouraged to engage with these ideas critically, and not as authoritative replacements for scientific consensus or spiritual doctrine.

No claim is made that the models, principles, or conclusions described herein are definitive or universally accepted. They are part of an ongoing conversation—a "war of thoughts"—that values courage over certainty and exploration over conclusion.

If this book unsettles what you've been taught to believe, that is not an accident. It is an invitation.

Don't miss out!

Visit the website below and you can sign up to receive emails whenever Sandeep Chavan publishes a new book. There's no charge and no obligation.

https://books2read.com/r/B-A-EPGPC-OKSDG

Did you love *Cracks in the Known*? Then you should read *The Future Knows No Lines*[1] by SANDEEP CHAVAN!

[2]

The Future Knows No Lines: Imagining a Borderless World is a thought-provoking exploration of what it means to live in a world without borders. Written by **Sandeep Chavan**, this book takes readers on an intellectual and imaginative journey, examining the possibilities, challenges, and implications of a unified global society.

The idea of a borderless world is sparked by a simple yet profound question posed by a young student: *"What if there were no borders? What would that world look like?"* From this starting point, the book dives deep into the complexities of governance, economy, culture, security, and education in a world where national boundaries are no longer relevant.

1. https://books2read.com/u/49KDGX

2. https://books2read.com/u/49KDGX

Through 17 captivating sessions that mimic interactive discussions, the author examines real-world examples of cooperation, such as the Schengen Area, global environmental agreements like the Paris Accord, and international aid coalitions. These case studies demonstrate how humanity has already begun taking steps toward reducing the significance of borders.

The book doesn't shy away from the challenges of such an idea. It critically analyzes the risks of resource exploitation, rising inequality, and cultural resistance, while also offering practical solutions, including global governance models, skill-based equity programs, and open migration policies. It explores emerging threats like cyber warfare, ideological clashes, and surveillance concerns, painting a balanced picture of both the opportunities and obstacles in a borderless world.

Education and mindset change emerge as central themes throughout the book. The author emphasizes the importance of teaching empathy, global citizenship, and critical thinking to future generations, empowering them to embrace collaboration over conflict. With technology bridging gaps and uniting people across continents, the potential for global learning and innovation is greater than ever before.

The Future Knows No Lines challenges readers to rethink their assumptions about borders and divisions. It is not merely an idealistic vision but a practical roadmap for building a more equitable, interconnected, and sustainable future. Through collaboration, education, and shared responsibility, the book inspires readers to imagine and work toward a world where humanity thrives as one.

This is more than a book—it's a call to action for everyone who dares to believe in the power of unity and the possibilities of a borderless future.

Read more at https://www.sandeepjchavan.com/.

Also by Sandeep Chavan

Echoes of War: The Russia-Ukraine Conflict and Its Global Ripples

Shattered Lives

The War Economy

Borders Redrawn

Arms and Allies

Collateral Damage

Justice in the Ashes

Somewhere in the Darkness of the Cosmos

The Blueprint of the Universe

The Cosmic Breath

The Eternal Sage Chronicles

The Sage and the Sovereign

The Family Pillar

The People's Flame

The Warrior's Soul

The Bridging Mentor

The Guiding Light

The Exodus
The Last Cycle: 2075

The Fundamental Theory of Life & Existence
Energy in Action: The Soul Demystified
The Active Mind
Energy in Connection
Death: The Disconnection
Subconsciousness: The Chosen Disconnection

The Restoration of Human Knowledge
Restoring Bhagavad Gita

The War of Thoughts
Cracks in the Known

Ultra Modern Life Philosophy
The Duality of Laughter and Tears
Take It Easy
Compare to Accept

Standalone
It's Not AI, It's AHI - Amplified Human Intelligence

The Decision Paradox
The Four Sapiens
Malicious Script of Indian Polity
The IIT Legacy & Global Impact
The IIT Legacy & Global Impact
You are just a Version of Your Original
Win the Game You Didn't Choose
Always Keep Your Bags Packed
Reveal to Shield - Learn the Game of Facades
Engage Beyond Elections
Culture, Identity & Change: The Evolution of Indian Society
The Triangular Dynamics
Never Trust Without Doubt
If You Can't Change the Country, Leave the Country
Live Fearless with Fear
India: The Investment Magnet
Realignment
Powering a Billion Dreams
Universal Skill
Extension: From Brush to Algorithm
Human Side of Einstein
The Future Knows No Lines
Let's Bring the Generations to the Table
Unified Energy Dynamics
A Beat That Never Dies
Strategic Horizons
Right From Scratch
The Forgotten Innovators
Physics Fixing
The Savior Illusion
The Selfish Side
The Age of the Weakest Sapiens
Neural Wars

The Absorption: There Is No Death
Strange Therapist

Watch for more at https://www.sandeepjchavan.com/.

www.ingramcontent.com/pod-product-compliance
Lightning Source LLC
LaVergne TN
LVHW090315210926
848480LV00001B/260